NITRATES UPDATED
Current Use in Angina, Ischemia, Infarction and Failure

Disclaimer

Although every effort has been made to ensure that the indications and doses for the various drugs are correct, the ultimate responsibility for the correct use of drugs lies with the prescribing physician. Likewise the institutions with which the author is associated have no direct nor indirect responsibility for the contents of this book nor for the mode in which any drugs in this book are used.

NITRATES UPDATED
Current Use in Angina, Ischemia, Infarction and Failure

Edited by:

Udho Thadani, MBBS, MRCP, FRCPC, FACC

Professor of Medicine,
Vice Chief Cardiology and
Director of Clinical Research
University of Oklahoma,
Health Sciences Center
Oklahoma City, Ok 73104

Lionel H. Opie, M.D., Ph.D.

Professor of Medicine
University of Cape Town
Cape Town, South Africa
Visiting Professor
Division of Cardiology
Stanford University Medical Center
Stanford, California USA

KLUWER ACADEMIC PUBLISHERS
BOSTON • DORDRECHT • LONDON

Distributors for North America:
Kluwer Academic Publishers
101 Philip Drive
Assinippi Park
Norwell, Massachusetts 02061 USA

Distributors for all other countries:
Kluwer Academic Publishers Group
Distribution Centre
Post Office Box 322
3300 AH Dordrecht, THE NETHERLANDS

Library of Congress Cataloging-in-Publication Data

Nitrates updated: current use in angina, ischemia, infarction and failure/edited by Udho Thadani, Lionel H. Opie.
p. cm.
Includes index.
ISBN 0-7923-3466-3 (alk. paper)
1. Nitrates—Therapeutic use. 2. Heart—Diseases—Chemotherapy. I. Thadani U, & Opie, Lionel H.
[DNLM: 1. Nitrates—therapeutic use. 2. Cardiovascular Diseases—drug therapy. QV 156 N7319 1995]
RC684.N56N586 1995
616.1'2061—dc20
DNLM/DLC
for Library of Congress 95-14210
CIP

Printed on acid-free paper.

Printed in the United States of America

CONTENTS

CONTRIBUTING AUTHORS

Johnathan Abrams
Division of Cardiology
University of New Mexico
School of Medicine
Albuquerque, New Mexico 87131

Johan Ahlner
Department of Clinical Pharmacology
University Hospital
S-581 85 Linkoping, Sweden

E. Bassenge
Institute of Applied Physiology
Hermann Herder Str. 7
D-79104 Freiburg, Germany

John Anthony Bauer
Department of Pharmaceutics
School of Pharmacy
State University of New York at Buffalo
Buffalo, New York 14260, USA

Marc G. Bogaert
Heymans Institute of Pharmacology
University of Gent Medical School
De Pintelaan 185
B-9000 Gent, Belgium

Jocelyn Dupuis
Montreal Heart Institute
5000 Belanger Street
E. Montreal
Quebec, Canada H1T 1C8

K. Fox
Royal Brompton Hospital
Sydney Street
London SW3 6NP UK

Ho-Leung Fung
Department of Pharmaceutics
School of Pharmacy
State University of New York at Buffalo
Buffalo, NY 14260, USA

B.I. Jugdutt
Division of Cardiology
2C2.43 Walter Mackenzie Health Sciences Centre
University of Alberta
Edmonton, Alberta, Canada T6G 2R7

Raymond J. Lipicky
FDA
Division of Cardiorenal Drug Products
5600 Fishers Lane
Room 16B-45/HFD 110
Rockville, MD 20857

D. Mulcahy
Royal Brompton Hospital
Sydney Street
London SW3 6NP UK

Lionel H. Opie
University of Cape Town
Heart Research Unit
Observatory 7925
Cape Town, South Africa

Henry Purcell
Royal Brompton Hospital
Sydney Street
London SW3 6NP UK

Udho Thadani
Cardiology Section
University of Oklahoma HSC
920 S.L. Young Blvd.
5SP-300
Oklahoma City, OK 73104

Kristina E. Torfgard
Department of Clinical Pharmacology
University Hospital
S-581 85 Linkoping, Sweden

PREFACE

Nitrates remain among the most commonly used therapeutic agents in cardiology. That they can relieve pain in angina pectoris has been known at least since 1867, when Lauder Brunton described their clinical use. Now, almost 130 years later, we are constantly learning more about their optimal use in anginal syndromes. For example, there is no doubt that nitrate tolerance occurs, but equally no doubt that certain drug dosages help to attenuate such tolerance. Thus, our understanding of nitrate tolerance has been taken further by the advent of new nitrate preparations and by studies at a basic science level.

The current intense interest in nitrates has been rekindled by the growing emphasis on the role of the endothelium in the production of nitric oxide, a basic cardiovascular regulator. If nitric oxide is vasodilator and protects the vascular endothelium, then nitrates should also have similar properties. Hence this book starts off with the role of the endothelium and nitrates in vasodilation. Next, the mechanisms of action of nitrates are considered with emphasis on the role of cyclic GMP. The chief limitation to the use of nitrates lies in nitrate tolerance. After carefully looking at all the evidence, the authors of the third chapter conclude that the mechanism is likely to be multifactorial. The mechanism responsible for the tolerance still remains unknown. In vitro tolerance may be due to enhanced production of vascular superoxide [1]. Regarding pharmacokinetics, it is proposed that for several reasons including the appearance of active metabolites, and attenuation of nitrate effects with time, the relationship between the plasma concentrations of nitrates and their therapeutic effects is very complex. Thus blood nitrate levels are never used by clinicians.

It is sobering to realise that none of the dosing regimens of any long-acting oral nitrate provides 24 hour anti-anginal and anti-ischemic effects. Furthermore, although it is intuitively attractive to add a β-blocker or calcium channel blocker to nitrate therapy, there are no objective studies to prove that such combinations actually give added benefits. Thus, much remains to be established. Regarding nitrate patches, the current opinion is that continuous therapy produces tolerance and is ineffective. Removing patches at night minimises tolerance, but introduces the risk of rebound nocturnal angina. In the case of unstable angina, the use of intravenous nitrates is standard, and is added to aspirin and intravenous heparin. In myocardial infarction, titrated intravenous nitrates are often used to help control pain and to limit ischemic injury, yet nitrates do not improve survival [2,3]. In silent ischemia, nitrates are safe and reduce the silent episodes. Nonetheless there is no specific proof that the treatment of silent ischemia by nitrates or any other drugs improves the outcome. In contrast, revascularisation reduces hard end points such as mortality, myocardial infarction and hospital admissions, according to the ACIP Study 4. Thus, the role of nitrates and other anti-ischemic drugs is to relieve pain as part of an overall management strategy. In congestive heart failure, nitrates are commonly used to help relieve dyspnea. Long-term combination of nitrates with hydralazine is beneficial. To lessen the risk of tolerance, it makes sense to include a nightly nitrate-free interval. Regarding practical problems in the use of nitrates, the only proven strategy to prevent tolerance is to employ a nitrate-free dosing interval and to avoid sustained nitrate delivery.

These important messages are here conveyed by current opinion leaders in the area of nitrate action, mechanism, and therapeutic use.

The original articles on which this book is based appeared as a series in the journal, Cardiovascular Drugs and Therapy. We are grateful to the authors and publishers to allow publication in this book.

We believe that our book will appeal not only to cardiologists and physicians who need guidance on nitrate therapy, but to basic workers who are interested in the mechanisms of the current problems that must be fully understood in order to achieve optimal nitrate therapy.

U Thadani and LH Opie,
Editors

References

1. Munzel T, Sayegh H, Freeman BA et al. Evidence for enhanced vascular super oxide. Anion production in nitrate tolerance: a novel mechanism of tolerance and cross tolerance. *J Clin Invest* 1995;95:187–194.
2. GISSI-3. Gruppo Italiano per lo Studio della Sopravvivenza nell'Infarto Miocardico, effects of lisinopril and transderman glyceryl ttinitrate singly and together on a 6 week mortality and ventricular function after acute myocardial infarction. *Lancet* 1994;343:1115–22.
3. ISIS-4 (Fourth International Study of Infarct Survival) Collaborative Group. ISIS4: A randomised factorial trial assessing early oral captopril, oral mononitrate, and intravenous magnesium sulphate in 58050 patients with suspected acute myocardial infarction. *Lancet* 1995;345:669–85.
4. Rogers WJ, Bourassa MG, Andrews TC. Asymptomatic cardiac ischemia pilot (ACIP) Study: outcome at 1 year for patients with asymptomatic cardiac ischemia randomised to medical therapy or revascularisation. *J Am Coll Cardiol* 1995;26:595–605.

Coronary Vasomotor Responses: Role of Endothelium and Nitrovasodilators

E. Bassenge

Summary. The endogenous nitrovasodilator endothelium-derived nitric oxide (EDNO) is continuously synthetized enzymatically by NO synthase from L-arginine and is released from endothelial cells. Enhanced, superimposed EDNO release can be stimulated by various local and circulating factors, such as bradykinin, ATP, etc., but also most importantly by viscous drag-induced shear stress of the blood-stream acting on the endothelial lining. Thus luminal release suppresses leukocyte adhesion (expression of adhesion molecules), platelet activation, platelet adhesion, and platelet aggregation, and abluminal release counteracts myogenic and neurogenic coronary constrictor tone, thereby increasing myocardial perfusion and dilating large coronary artery calibers. Thus endothelial impairment and denudation (hypercholesterolemia, atheromatosis, balloon catheter interventions) favor excessive constrictor tone and myocardial ischemia. Under these conditions EDNO can be supplemented by compounds (e.g., nitroglycerin, isosorbide dinitrate) converted by biological systems into NO. In addition, it can be supplemented by compounds that even spontaneously release NO (e.g., sydnonimines such as SIN-1 and sodium nitroprusside). EDNO and exogenously supplemented NO stimulate soluble guanylyl cyclase, increase cGMP levels, and bring about vascular relaxation, particularly in those still compliant sections in which EDNO production is impaired and cGMP levels are thus diminished. Exogenous nitrovasodilators are preferentially converted (in the presence of cysteine) enzymatically in large coronary arteries, improving coronary conductance, and in the venous bed (preload reduction), resulting in an improved O_2 supply/demand ratio. During chronic, continuous application, neurohormonal counterregulation and diminished enzymatic biotransformation into NO may reduce their effectiveness, resulting in tolerance, particularly in the most sensitive vascular sections, such as veins and coronary arteries. This drawback can be overcome by spontaneously applying NO-releasing compounds, intermittent therapy, or intermittent interposition of other vasodilator principles.

Key Words. **endothelium derived NO, nitrovasodilators, nitrate tolerance, receptor dependent and independent NO release, nitric oxide**

Physiological Interactions

Multiple factors interact in the regulation of coronary tone

Coronary tone, and thus myocardial perfusion and oxygen supply, are regulated by metabolic, myogenic, humoral-hormonal, neuronal, and endothelial factors. The importance of autacoids such as endothelium-derived nitric oxide (EDNO), prostacyclin (PGI_2), platelet activating factor (PAF), and endothelin (ET), synthesized and released from the endothelial lining, for vascular control and hemodynamic homeostasis has been amply demonstrated within the last two decades [1,2]. Their significance under pathophysiologic conditions, such as coronary artery disease, hypercholesterolemia, hypertension, diabetes, and heart failure, has only become evident during the last years.

Coronary artery disease is usually limited to the large epicardial arteries, the tone of which is predominantly regulated by endothelial, in addition to humoral-neurohormonal, factors. Under exercise sympathetic effects (norepinephrine from nerve endings and circulating catecholamines) may gain a certain significance [2]. As indicated in Figure 1, myogenic factors and also metabolic factors,are less active in epicardial conduit arteries, due to the large diffusion distance between the adjacent myocardium (releasing catabolites) and the vasculature.

As outlined in Figure 2, however, endothelial factors such as EDNO continuously synthesized and released, both without stimulation (so-called unstimulated or *basal NO release*) and with stimulation (*stimulated release* which is elicited by local and circulating agonists such as bradykinin, serotonin, norepinephrine, etc and by viscous drag-induced shear stress from the perfusing blood resulting in flow-dependent dilation), exhibit pronounced vasomotor effects on coronary tone both in large conductance arteries and in the coronary microvasculature. The tone of corononary arterioles predominantly determines myocardial perfusion and its distribution. This microvascular tone is

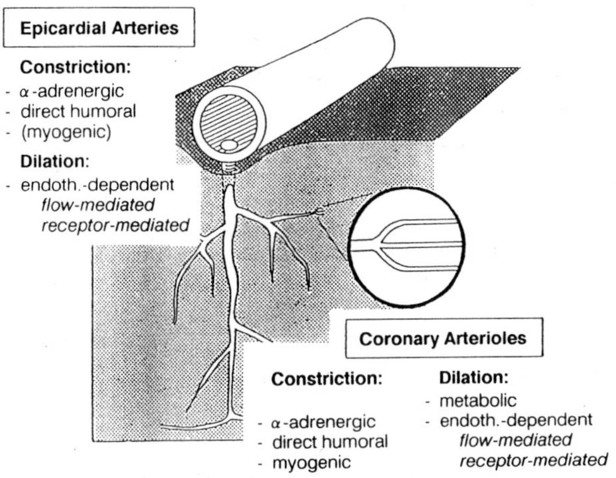

Fig. 1. Regulatin of coronary blood flow. Simplified scheme representing regulation of coronary tone and myocardial perfusion in large conduit arteries as compared to coronary resistance vessels. Several constrictor and dilator mechanisms interact differently in the two vascular sections, as described in the text. There is a gradual shift in the importance of metabolic factors from large arteries (almost absent) towards smaller arteries, smallest arteries, and arterioles (maximal).

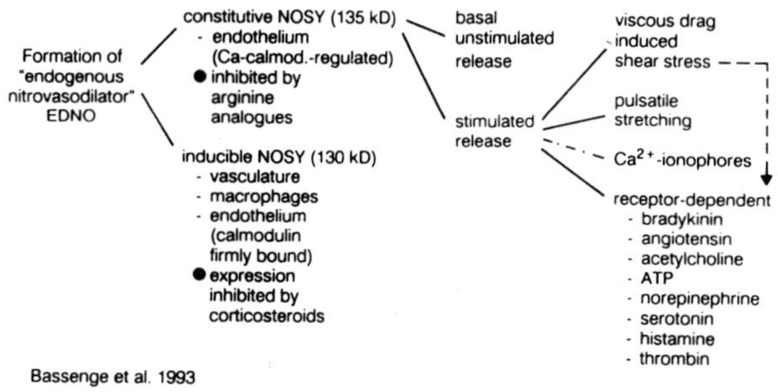

Fig. 2. Formation of the endogenous nitrovasodilator nitric oxide by different constitutive (endothelial, cerebral) and inducible (vascular smooth muscle, macrophages, endothelium) isoforms of NO synthase. Receptor-dependent NO release is potentiated by shear-stress-dependent NO release (for further details, see text), NOSY = nitric oxide synthase; EDNO = endogenous nitric oxide.

adjusted primarily by superimposed metabolic factors and, in a less pronounced manner by hormonal and neuronal factors in addition to myogenic effects (Figure 1). These myogenic effects also seem to depend, at least in part, on an unimpaired endothelial function [3]. Under physiologic conditions the continuous dilator action of endothelial autacoids compensates for the continuous constrictor effect of norepinephrine released from nerve endings [4]. In the absence of a functionally intact endothelial lining and autacoid production in coronary artery disease, both shear stress-dependent NO release (which elicits flow-dependent dilator responses and adjusts the conductance of the large feed arteries in an "ascending manner") and the concomitant myogenic responses are substantially affected: While the flow-dependent, NO-mediated dilator responses are diminished [2,5], the myogenic constrictor responses are enhanced [6], which also results in an augmented autoregulatory response [7].

Histopathological alterations in the coronary microvasculature, which are associated with "microvascular angina," are rarely observed in patients. In microvascular angina the maximal capacity to dilate—in the absence of any angiographical alterations in the conductance arteries—is severely reduced [8]. Even in the course of normal aging, the capacity to synthesize and release endothelial autacoids seems to be progressively reduced, thus continuously limiting the maximal capacity to dilate, even in the absence of significant pathological alterations at higher age [9]. A similar reduction, both in the maximum capacity to dilate (and thus in coronary reserve) and in resting flow, can be brought about by the administration of specific inhibitors of endothelial NO synthesis in test animals using arginine analogues such as L-N-monomethylarginine (L-NMMA) or similar analogues (see below) [10,11]. This reduction cannot be normalized by additional, acute application of any other coronary vasodilator.

The endothelial lining modulates coronary tone not only by the basal and/or stimulated synthesis and subsequent release of endothelial autacoids, but also by the uptake, storage, and metabolism of numerous vasoactive circulating compounds, such as serotonin, norepinephrine, etc. Furthermore, EDNO also modifies transmitter release from nerve endings, thereby modulating and reducing alpha-adrenergic constrictor stimuli. In certain vascular areas NO is released from nitroxydergic nerve endings, thus increasing tissue perfusion (e.g., intestinal blood flow) [12]. Finally, in the brain locally synthesized NO suppresses the continuous sympathetic activity [13]. Therefore, inhibitors of NO synthesis, administered intracisternally, stimulate sympathetic activity and thus enhance alpha-adrenergic coronary constriction [13] and increase blood pressure.

As depicted in Figure 3, there is a broad spectrum of different factors and drugs that stimulate (both in a receptor-dependent or -independent manner) the production and release of endothelial autacoids (NO, prostacyclin, endothelin [14])-like platelet factors, hemostasis factors (such as thrombin [14]), vasoac-

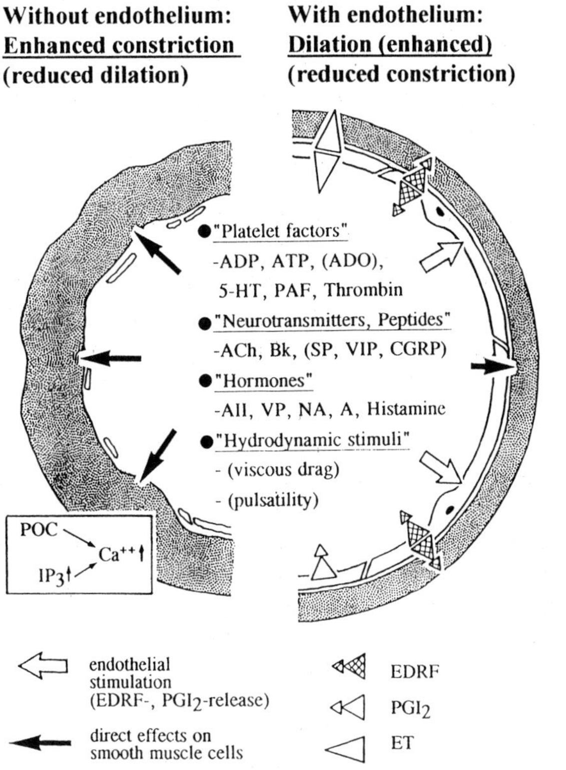

Fig. 3. Simplified scheme of endothelium (E)-mediated control of coronary tone. In the absence of a functionally intact endothelial lining (e.g., balloon catheter-induced endothelial denudation, coronary artery disease), endogenous and exogenous compounds, and factors affect vascular tone directly, predominantly resulting in usually inositol-trisphosphate (IP3) mediated constriction (left-hand side). POC = potential operated calcium channels. The intact endothelium has a decisive modulatory role for the various stimulatory compounds and factors eliciting usually NO-mediated, superimposed dilator responses (right-hand side). Therefore, the combined net response of direct and E-mediated effects is usually dilation or attenuated constriction. The pathophysiological significance of E-dependent, thrombin-induced endothelin release and its inhibition by EDRF or nitrovasodilators has not been clarified yet [14]. Prostacyclin release is predominantly towards the luminal side, whereas EDRF/NO release is similar on both the luminal and the abluminal side. Thus, the actual level of tone is determined by the net effect resulting from both E-mediated dilator effects and simultaneous direct E-independent constrictor effects (by NA, VP, AII, histamine) or direct, E-independent dilation (beta-adrenoceptor or cAMP mediated). Substances/factors in parentheses only elicit vasodilation. 5-HT = 5-hydroxytryptamine (serotonin); ADO = adenosine; PAF = platelet activating factor; ACh = acetylcholine; Bk = bradykinin; SP = substance P; VIP = vasoactive intestinal polypeptide; CGRP = calcitonin generelated peptide; AII = angiotensin II; VP = vasopressin; NA = noradrenaline; A = adrenaline; ET = endothelin.

tive peptides, neurotransmitters, various hormones, Ca^{2+} ionophores, and in particular, hydrodynamically induced viscous drag of the blood and pulsatile stretching of the endothelial lining [15]. These stimuli trigger an intracellular Ca^{2+} signal, which in turn enhances the activity of NO synthase. Viscous drag apparently acts on endothelial K^+ channels [16], causing hyperpolarization, transmembrance Ca^{2+} influx, and an increase in intracellular Ca^{2+}, leading to an enhanced release of PGI_2 and EDNO due to stimulation of the Ca^{2+}-dependent phospholipase A2 and the constitutive endothelial NO synthase. In addition, stimulation of transmembrane Ca^{2+} influx [17] as a potential trigger signal for NO synthesis through nonspecific cationic channels has been demonstrated. The pulsatile stretching of the endothelial lining adds to this mechanically stimulated NO release [15]. Chronically applied pulsatile stretch and/or viscous drag induce enhanced expression of the key enzyme in endothelial NO production, namely, the endothelial 135-kD Ca-calmodulin-dependent NO synthase [18,19], which has recently been identified and cloned [20]. Viscous draginduced shear stress and pulsatile stretching of the endothelial lining also potentiate agonist-induced, receptor-dependent stimulation of NO release [21].

All vasoactive compounds and drugs, therefore, modulate endothelial NO synthesis and release in either a receptor-dependent manner, as indicated in Figure 2 (far right-hand side), or in a flow-dependent, shear-stress-mediated manner [21].

Isoforms of NO synthase

The "endogenous nitrate" EDNO is continuously synthesized from arginine (Figure 4) in the presence of various cofactors, such as NADPH, tetrahydrobiopterin, flavin adenine dinucleotide (FAD), and flavin adenine mononucleotide (FMN), by a recently discovered and cloned constitutive enzyme, called endothelial NO synthase. At the end of this reaction (in which L-arginine is transported into the endothelial cell by a specific carrier and acts as a precursor), one citrulline molecule is formed and one NO radical is split off (Figure 4). This radical leaves the endothelial lining either intraluminally to inhibit leukocyte adhesion suppressed expression of adhesion molecules [22–24] and platelet adhesion and activation, or abluminally to modulate vascular tone [20] or vascular growth processes [25]. This enzyme has a surprising amount of homology with P-450 cytochrome oxidase and exists in at least three different isoforms: Isoform I (MW 160kD) is a constitutive enzyme in the *brain* responsible for NO production

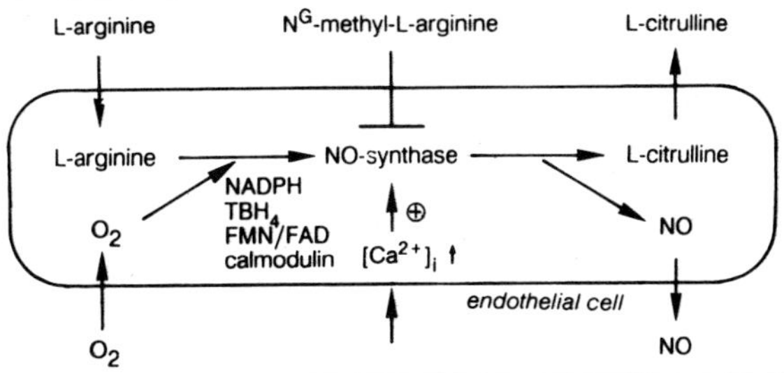

Fig. 4. *Biosynthesis and release of the endogenous nitrate endothelium derived nitric oxide (EDNO) from the precursor L-arginine in the presence of molecular oxygen (O_2). The end products are nitric oxide (NO) and L-citrulline. The reaction is catalyzed by endothelial NO synthase, which is activated by intracellular rises in (Ca^{2+}) in the presence of several cofactors such as NADPH, tetrahydrobiopterin (TBH_4), flavin, adenine, mono- and di-nucleotide (FMN, FAD). NG monomethyl arginine inhibits NO synthase.*

and serving as a transmitter substance in neuronal cells that is regulated by Ca^{2+} and calmodulin. Isoform II (MW 130 kD) is an inducible enzyme expressed in *macrophages* [26] and is responsible for the NO-induced cytotoxic effects of these cells [27]. A similar isoform can be induced in septic shock by cytokines in the vasculature, as indicated in Figure 5, and is responsible for excessive vasodilation. Its activity is not regulated by Ca^{2+} because calmodulin is already firmly bound to the enzyme. The expression of this enzyme is inhibited by corticosteroids [28], which may explain their inhibitory effect on inflammation-induced hyperemia.

Isoform III (MW 135 kD) is the decisive constitutive enzyme responsible for continuous EDNO production in *endotheial cells* and is also regulated by Ca calmodulin [20]. It was recently characterized and cloned, and antibodies have been developed. An *inducible* isoform (similar to type II, see below) can additionally be expressed in endothelial cells. Its pathophysiological importance lies in the expression of inducible isoforms in the vasculature by cytokines and similar factors in various diseases, e.g., in septic shock. They are responsible for a therapy-resistant, excessive vasodilation and serious hypotension. This can be beneficially affected by administration of arginine analogues acting as specific inhibitors of vascular NO synthesis [29,30]. From these findings it becomes evident that states of excessive coronary vasocon-strictor tone, associated with a limited capacity to dilate and with restricted myocardial perfusion, cannot simply be therapeutically compensated for by stimulating all of the different NO synthases in a rather unspecific manner (Figure 5) in the vessel

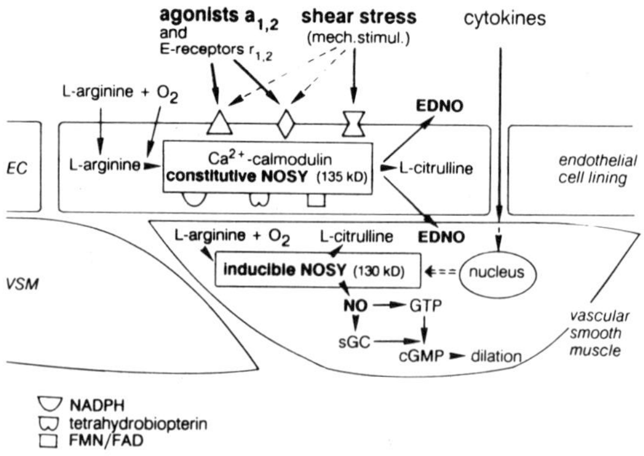

Fig. 5. Various isoforms of NO synthase (NOSY) in the vessel wall. Constitutive, endothelial NO synthase, is regulated by Ca^{2+} calmodulin and is stimulated by specific (receptor-dependent) agonists and/or by flow-induced shear stress. Under pathophysiologic conditions, a nonregulated, inducible NO synthase is expressed following cytokine stimulation in vascular smooth muscle and elicits excessive, uncontrolled dilator responses, which can be blocked by inhibitors of NO synthase. E = endothelin; EC = endothelial cell; sGC = soluble guanylyl cyclase.

wall instead of selectively stimulating deficient, well-regulated endothelial NO production: Adequate therapeutic NO supplementation by *exogenous* nitrovasodilators or by stimulating appropriately endogenous EDNO production has to be well defined and restricted to specific coronary sections, characterized by impaired EDNO production (e.g., endothelium-denuded or atherosclerotic, but still compliant, sections, such as eccentric stenoses characterized by near-normal wall segments observed in part of the vessel circumference) [31].

Stimulation of *endogenous* No synthesis by enhanced oral and/or i.v. arginine supplementation has been successfully applied in various states associated with restricted coronary dilator reserve, such as hypercholesterolemia, atheromatosis, and in the post-transplantation state [32–34]. However, in many other diseases, such a limitation in arginine supply and/or transmembrane arginine transport does not seem to be a decisive parameter in the limitation of adequate dilator responses.

Pharmacological Interventions

Supplementation of impaired NO synthesis by exogenous NO donors

When exogenous NO donors (nitrovasodilators, such as organic nitrates, sydnonimines, and furoxan derivatives) are employed in the treatment of myocar-

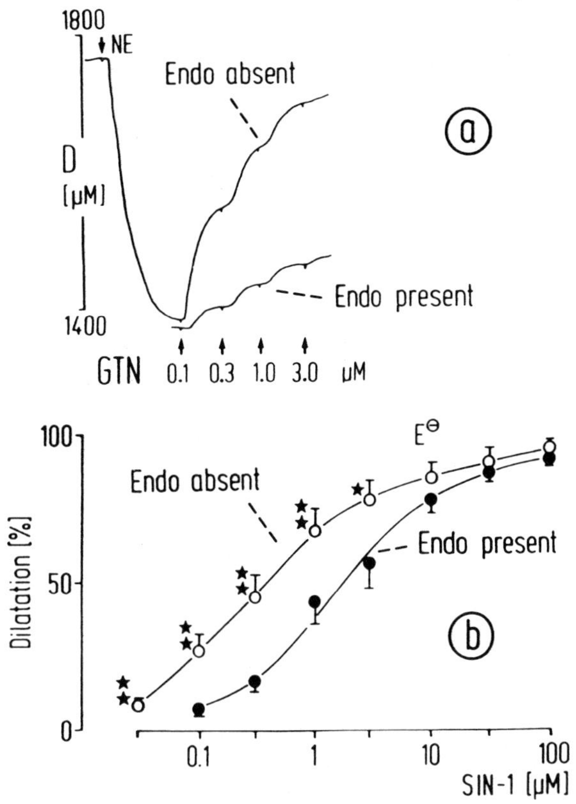

Fig. 6. *A: Increased sensitivity of vascular segments perfused in vitro with and without endothelium (Endo) during cumulative administration of nitroglycerin (GTN) in the perfusate. D = outer diameter of the segments; NE = noradrenaline, 3 × 10 − 7M [37]. B: Cumulative concentration-response curve to NO-releasing SIN-1 with endothelium (control). Impairment of formation of endogenous nitric oxide (EDNO) with NG-nitro-L-arginine (30 μM) significantly shifts the concentration-response curve to the left (NO synthesis inhibited). Dilatation is expressed as the percent of initial steady-state contraction induced by norepinephrine (0.11 = 0.04 μM before the addition of SIN-1 *p < 0.05, ++P < 0.01) [38].*

dial ischemia, one can observe a very favorable distribution of rather selectively localized dilator responses within the vascular bed, resulting in an almost unique (and therapeutically most appropriate) pattern of differential drug actions.

The smallest effective nitrate concentrations result in venodilation, reduction in cardiac preload, and myocardial oxygen consumption, as well as dilation of the large coronary conductance arteries [35]. This is due to the fact that the vasculature of these specific sections has the greatest capacity to metabolize nitrates to NO, whereas in the arterioles and other microvessels, the activity of these specific enzymes is

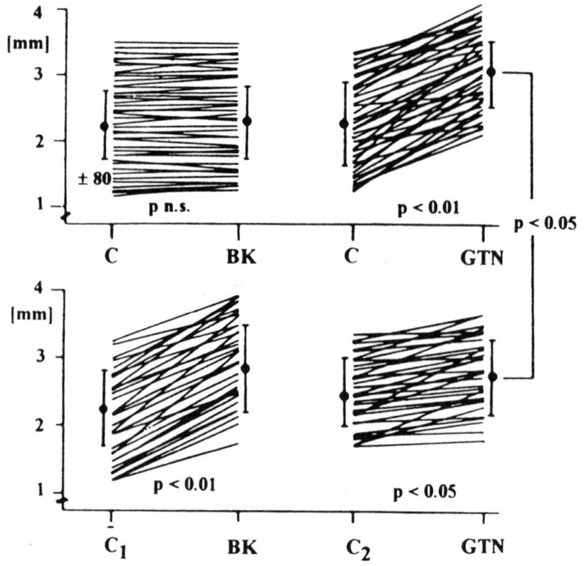

Fig. 7. *Top panel: Demonstration of augmented nitroglycerin (TNG) induced dilator response in coronary arteries of patients with loss of endothelium-dependent dilation following intracoronary administration of bradykinin (BK). Bottom panels: In coronary sections with intact endothelium-dependent dilator responses to BK, the nitroglycerin-induced dilator response is much less pronounced (bottom right). (Modified from Raffenbeul et al. [40], with permission.)*

almost negligible [36]. In ischemic heart disease, the conductance arteries are usually affected most severely by an impairment of endothelial function and arteriosclerosis. The restricted conduction of these stenotic sections needs to be therapeutically augmented to improve the oxygen supply to the heart.

Finally, with regard to differential nitrovasodilator effects in different sections, it has been shown (Figure 6a and 6b) both in vitro [37–39] as well as in clinical studies (Figure 7) [40], that *exogenous* supplementation by NO donors results in preferential dilation of segments with impaired endothelial function (e.g., atheromatotic, yet *still compliant*), in which endogenous EDNO production, and thus vascular cGMP levels and continuous dilator responses, are particularly reduced [41]. Supplementation by exogenous NO donors such as the classic organic nitrates [42], in addition to sydnonimine [43,44] and furoxan derivatives [45], therefore effectively and preferentially augments cGMP levels, especially in the affected—and consequently autacoid—and cGMP-deficient sections. In these sections the superimposed stimulation of soluble guanylyl cyclase (sGC) by NO results in particularly effective cGMP production in the vasculature! This, in turn, improves both coronary conductance and precollateral perfusion pressure. Furthermore, in the presence of a nitrovasodilator-induced reduction in left ventricular filling and end-diastolic pressure, this will result in unloading of the subendocardial segments and passive improvement of subendocardial perfusion, preventing ischemic damage in the endangered ventricular sections.

Only if effective concentrations of the nitrovasodilators are substantially raised (about 10 times) will significant effects on peripheral resistance vessels and blood pressure become apparent, thus eliciting unwanted, pronounced drops in coronary perfusion pressure in the course of antiischemic therapy by nitrovasodilators. Finally, if effective nitrovasodilator concentrations are still raised by another decade, only then coronary arteriolar dilator effects become more significant (due to the limited nitrate conversion to NO in these sections) [36]. Thus, in the presence of specifically localized stenoses even coronary steal phenomena may be evoked.

During tolerance a shift towards a more unfavorable spectrum of nitrovasodilator actions is observed

During chronic continuous nitrovasodilator therapy (e.g., during unstable angina or following myocardial infarction) the vasculature becomes more or less unresponsive (tolerant) to organic nitrates in specific sections [46]. Tolerance first becomes evident in the venous bed and in large coronary arteries while peripheral resistance vessels remain responsive. Tolerance is a multifactorial phenomenon, which occurs during continuous, high-dose nitrate therapy and rarely in the course of intermittent (low-dose) therapy [47–49]. Under clinical conditions neurohormonal counterregulation [50] accounts for most of the efficiency loss ("pseudotolerance"; see Figure 8. The cyclic ADP-ribose induced rise in intracellular Ca^{2+} [51] has not yet been demonstrated in the vasculature). With higher concentrations an impairment of an enzyme-dependent biotransformation into the active NO-releasing compound takes place, as shown in Figure 8. Under clinical conditions this state may be nicely reflected by an impaired NO-induced rise in platelet cGMP levels, which is well correlated with a reduced coronary dilator response [52]. At the highest concentrations (pharmacological in vitro experiment in the millimolar range), desensitization of soluble guanylyl cyclase has been demonstrated

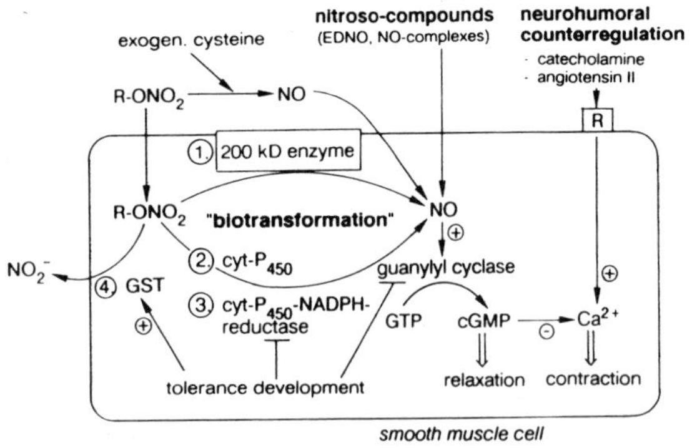

Fig. 8. *Activation of soluble guanylyl cyclase initiates vascular relaxation in a cGMP-dependent manner. The enzyme is activated by NO derived from several possible sources. Exogenous nitrates are converted into NO by enzymatic conversion. Exogenous/endogenous nitroso compounds, such as endogenous NO (EDNO), dinitrosyl iron complexes, and sodium nitroprusside spontaneously release NO. Such biotransformation may become inhibited, leading to drug tolerance. This limitation may result from the downregulation or partial inactivation of enzymes 1, 2, and 3. Activity of glutathion-S transferase (GST, 4) leads to transformation of NO into vasoinactive nitrite ions. The cGMP-induced decrease in intracellular Ca^{2+} may be partly counteracted by a cGMP-dependent activation of the ADP-ribosyl cyclase pathway, forming cyclic ADP-ribose (c-ADPR), which acts as an intracellular messenger to increase intracellular Ca^{2+}, for instance, in cardiac muscle [51]. The importance of this additional pathway in the coronary vasculature cannot be judged at present. (1) 20 kD NO-generating enzyme from plasma membrane fraction; (2) microsomal cytochrome P-450; (3) microsomal cytochrome-P-450-reductase; (4) microsomal glutathion-S-transferase (GST).*

[53,54]. During tolerance under clinical dosages of nitrates (nanomolar range) the action of endogenous EDNO stimulated by agonists such as acetylcholine or increased viscous drag is not reduced [46].

To overcome tolerance concentrations of the nitrates have to be progressively increased during the therapeutic period, thereby enhancing the tolerance-resistant, unwanted side effects, such as substantial drops in blood pressure (and thus in coronary perfusion pressure!) associated with a reflex increase of heart rate and thus in myocardial O_2 consumption. Under such conditions the potential induction of intracoronary steal phenomena becomes more likely, while cardiac preload and coronary conductance cease to be beneficially affected. In conclusion, tolerance causes a shift towards unfavorable side effects, rendering therapeutic interventions not only ineffective but even potentially harmful.

Differential dilator actions of nitrates in specific vascular beds

The reason for the preferential dilator action of organic nitrates on large arteries and on the venous system originates from the different capacity of a preceding, enzyme-dependent metabolic conversion from the organic nitrate into an active, NO-radical-containing compound, probably a nitrosothiol, which stimulates soluble guanylyl cyclase. With the exception of sydnonimines (which spontaneously release NO during hydrolysis), and probably partly of sodium nitroprusside [55], this conversion of nitrovasodilators to NO depends on several, not yet completely identified, enzymes, which are mainly present in large arteries and veins, but rarely in any of the microvasculature [36,55], rendering the latter rather insensitive to organic nitrates but not to exogenous NO, or NO spontaneously released from the sydnonimine [44] sodium nitroprusside.

Nitrates do not only dilate selectively large conductance coronary arteries but dilate more or less all large conduit arteries. However, with regard to the significance of an inadequate blood supply, particularly to the myocardium in the presence of stenoses and of a deficient dilator capacity, the therapeutically induced dilator response, especially of the atherosclerotic and stenotic sections of the coronary arteries, is particularly effective and thus of pathophysiological consequence.

Potential NO-Mediated Therapeutic Interventions

Significance of chronically stimulated EDNO release

In recent years a number of cardiovascular disease states have been identified that are associated with endothelial dysfunction [56] and impaired endothelial NO synthesis and release. They are summarized in Figure 9. These disease states are similar in that constrictor responses predominate in all of them. These enhanced constrictor responses due to impaired endothelium and NO-mediated dilator responses should be preferentially compensated for by therapeutic stimulation of endothelial NO synthase activity. At present no such specific stimulators have been identified. Yet, in some disease states of the heart, such as hypercholesterolemia, experimental atheromatosis, and in the post-transplantation state, associated endothelial dysfunction and autacoid release can be corrected by oral or intravenous arginine

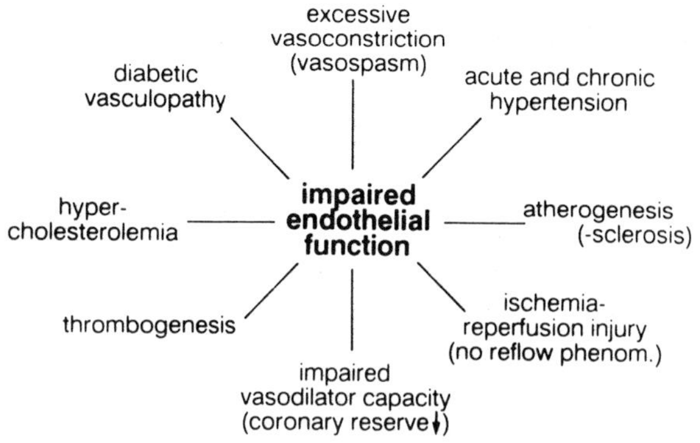

Fig. 9. *Impaired endothelial function is associated with a large number of cardiovascular disease. Whether the dysfunction is the cause for, or a consequence of, the disease has not been clarified yet (e.g., in hypertension). Usually the endothelial cells do not show major histological alterations, but autacoid formation and release is more or less affected. This can be attributed to changes in enzyme activity or expression (e.g., c-NOS) or to increased formation of O_2 radicals, which can act as scavengers for the NO radical. These various disease states may in a number of instances be associated with lipid droplets (atheromatosis, atherosclerosis), vacuolar degeneration, and augmentation of the endoplasmatic reticulum (effects of smoking), osmiophilic shrinkage, and fatty degeneration (diabetes), partial detachment, and loss of the endothelial lining (hypertension) [56].*

supplementation, which can normalize impaired endothelium-mediated dilator responses, even if the underlying disease remains present (e.g., hypercholesterolemia, atherosclerosis). This could easily be explained if the interesting, preliminary observation of reduced arginine plasma levels in hypercholesterolemia can be reproduced appropriately [57]. The finding of deficient NO production in atheromatosis and atherosclerosis is controversial: Recently enhanced NO production was found under these conditions and reduced NO release was explained by increased NO degradation through enhanced generation of oxygen radicals [58,59].

Another method of stimulating the expression of endothelial NO synthase employs chronic application of increased shear forces by increasing blood flow. This has been demonstrated to be effective and to induce augmented expression of mRNA for NO synthase and increased protein synthesis of NO synthase [18,19]. Thus, it is evident that a number of therapeutically rather ineffective coronary arteriolar dilators, such as dipyridamole and carbochromen (already at lower concentrations), as well as some Ca^{2+} antagonists of the dihydropyridine type (at higher concentrations), exhibit a dual dilator action: They stimulate, in addition to arteriolar dilation, a pronounced shear-stress-dependent NO synthesis and release in the preceding feeder arteries. Indeed, we were able to demonstrate that a substantial part of the Ca^{2+}-antagonist-induced dilator response of the large arteries (especially within the first 3–5 minutes after administration) results from the superimposed EDNO release from a functionally intact endothelial lining (Figure 10). This can be blocked and thus be quantified by keeping blood flow (and thus shear forces) experimentally constant [60] or by suppressing the concomitantly augmented EDNO production by specific inhibitors (L-NMMA).

In mid- and long-term animal experiments on new approaches to vascular protection during hypercholesterolemia, atherosclerosis, and hypertension, enhanced EDNO production has been shown to offer protection from leucocyte and monocyte infiltration, and atherosclerosis, and thus from myocardial ischemia. In addition, the increased platelet activation, adhesion, and aggregation is inhibited by EDNO, resulting in suppression of thromboembolic events, such as myocardial infarction or unstable angina, and in suppression of excessive vessel growth, e.g., following balloon catheter interventions.

Endothelial NO production can effectively be stimulated by chronic administration of ACE inhibitors. They inhibit (due to a favorable side effect)

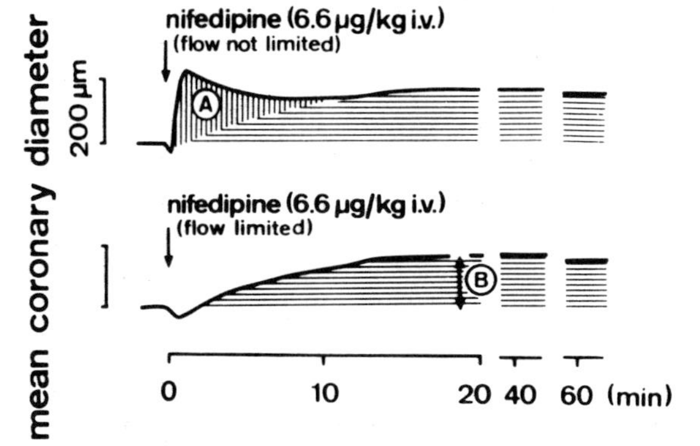

Fig. 10. Nifedipine-induced dilator effect on large coronary arteries comprises two components: a: Endogenous nitric oxide (EDNO) effect: Coronary arteriolar dilator response increases myocardial blood flow. This flow increase initiates EDNO release through increased shear stress resulting in immediate dilator response indicated by hatched area (A). b: Direct nifedipine effect: Arteriolar effect suppressed by flow-limiting stenosis keeps shear stress and EDNO production constant. The superimposed EDNO effect is completely suppressed, unmasking the isolated, direct nifedipine-induced dilator effect on large coronary arteries, which is much slower in onset and takes longer to develop (horizontally hatched area B) [60].

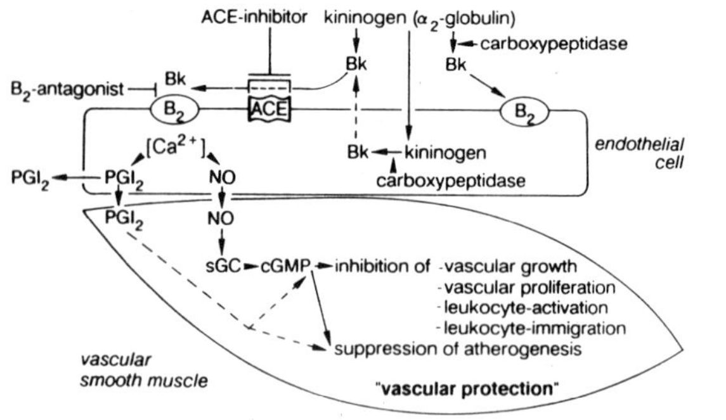

Fig. 11. ACE-inhibitor-induced accumulation of bradykinin (Bk) elicits vascular protection through NO generation. Activated carboxypeptidase splits from an α2-globulin the nonapeptide bradykinin. This peptide can either stimulate endothelial bradykinin 2 (B2) receptors to activate the constitutive endothelial NO synthase in an Ca^{2+}-dependent manner, or it can be degraded by the membrane-associated angiotensin converting enzyme (ACE) into vasoinactive peptides. This cleavage and inactivation can be suppressed by ACE inhibitors, resulting in augmented bradykinin concentrations, stimulation of NO synthase, and NO release. NO may suppress atherogenesis by a complex of multiple actions.

breakdown of bradykinin. The augmented bradykinin concentration, in turn, continuously stimulates, as shown in Figure 11, via bradykinin B2 receptors and subsequent increase in intracellular Ca^{2+}, NO synthesis and release from the endothelial lining, which results in significant vascular protection in animal studies [61]. Arginine supplementation in a variety of diseases has also been found to improve endothelial function and NO release [3,33,34]. The beneficial anti-atherosclerotic effect originates from a combination of various effects: suppression of platelet and leukocyte activation, neutrophil adherence, and neutrophil and monocyte immigration [62], and inhibition of expression of the leukocyte adhesion molecules CD 11 and CD 18 [24]. Leukocyte adhesion is an early step in the initiation of atherogenesis. NO has also been shown to inhibit both mitogenesis and excessive proliferation of vascular smooth muscle [63] and suppress the production of excessive matrix molecules, such as elastin and collagen. Consequently, gene therapy using cNOS gene transfection into the coronary bed may prove to be an effective strategy [64]. Thus, continuous NO release from the coronary endothelial lining has a multifactorial potential of protecting against atherosclerosis-associated restrictions of myocardial blood flow, which can only partly be mimicked by exogenously supplied NO donors.

References

1. Furchgott RF, Vanhoutte PM. Endothelium-derived relaxing and contracting factors. *FASEB J* 1989;3:2007–2018.
2. Bassenge E, Heusch G. Endothelial and neuro-humoral control of coronary blood flow in health and disease. *Rev Physiol Biochem Pharmacol* 1990;116:77–165.
3. Rubanyi GM, Freay AD, Kauser K, Johns A, Harder DR. Mechanoreception by the endothelium—mediators and mechanisms of pressure-induced and flow-induced vascular responses. *Blood Vessels* 1990;27:246–257.
4. Halbrugge T, Lutsch K, Thyen A, Graefe KH. Role of nitric oxide formation in the regulation of haemodynamics and the release of noradrenaline and adrenaline. *Naunyn Schmiedebergs Arch Pharmacol* 1991;344:720–727.
5. Zeiher AM, Drexler H, Wollschläger H, Just HJ. Modulation of coronary vasomotor tone in humans. Progressive endothelial dysfunction with different early stages of coronary atherosclerosis. *Circulation* 1991;83:391–401.
6. Pohl U, Lamontagne D, Bassenge E, Busse R. EDRF augments coronary conductivity through attenuation of myogenicmyogenic autoregulation (abstr). *Pflügers Arch* 1990;415(Suppl 1):R62.
7. Pohl U, Lamontagne D, Bassenge E, Busse R. Attenuation of coronary autoregulation in the isolated rabbit heart by endothelium derived nitric oxide. *Cardiovasc Res* 1994;28:414–419.
8. Quyyumi AA, Cannon RO, Panza JA, Diodati JG, Epstein SE. Endothelial dysfunction in patients with chest pain and normal coronary arteries. *Circulation* 1992;86:1864–1871.
9. Zeiher AM, Drexler H, Saurbier B, Just H. Endiothelium-mediated coronary blood flow modulations in humans: Effects of age, atherosclerosis, hypercholesterolemia, and hypertension, *J Clin Invest* 1993;92:652–662.
10. Huckstorf C, Zanzinger J, Bassenge E. Reduced nitric oxide formation causes coronary vasoconstriction and impaired dilator responses to endogenous agonists and hypoxia in dogs. *Naunyn-Schmiedebergs Arch Pharmacol* 1994;349:367–373.
11. Bassenge E. Endothelium-mediated regulation of coronary tone. *Basic Res Cardiol* 1991;86:69–76.
12. Toda N, Okamura T. Regulation by nitroxidergic nerve of arterial tone. *News Physiol Sci* 1992;7:148–152.
13. Sakuma I, Togashi H, Yoshioka M, et al. NG-methyl-L-arginine, an inhibitor of L-arginine-derived nitric oxide synthesis, stimulates renal sympathetic nerve activity in vivo—A role for nitric oxide in the central regulation of sympathetic tone. *Circ Res* 1992;70:607–611.
14. Boulanger C, Luscher TF. Release of endothelin from the porcine aorta—inhibition by endothelium-derived nitric oxide. *J Clin Invest* 1990;85:587–590.
15. Pohl U, Busse R, Kuon E, Bassenge E. Pulsatile perfusion stimulates the release of endothelial autacoids. *J Appl Cardiol* 1986;1:215–235.
16. Olesen SP. An electrophysiological study of microvascular permeability and its modulation by chemical mediators. *Acta Physiol Scand* 1989;136:7–28.
17. Shen J, Luscinskas FW, Connolly A, Dewey CF, Gimbrone MA. Fluid shear stress modulates cystolic free calcium in vascular endothelial cells. *Am J Physiol* 1992;262:C384–C390.
18. Nishida K, Harrison DG, Navas JP, et al. Molecular cloning and characterization of the constitutive bovine aortic endothelial cell nitric oxide synthase. *J Clin Invest* 1992;90:2092–2096.

19. Uematsu M, Navas JP, Nishida K, et al. Mechanisms of endothelial cell NO synthase induction by shear stress. *Circulation* 1993;88:I184.
20. Förstermann U, Pollock JS, Nakane M. Nitric oxide synthases in the cardiovascular system. *Trends Cardiovasc Med* 1993;3:104–110.
21. Hecker M, Mülsch A, Bassenge E, Busse R. Vasoconstriction and increased flow: Two principal mechanisms of shear stress-dependent endothelial autacoid release. *Am J Physiol* 1993;265:H828–H833.
22. Lefer AM, Siegfried MR, Ma X. Protection of ischemiareperfusion injury by sydnonimine NO donors via inhibition of neutrophil-endothelium interaction. *J Cardiovasc Pharmacol* 1993;22(Suppl):S27–S33.
23. Weyrich AS, Buerke M, Albertine KH, Lefer AM. Time course of endothelial adhesion molecule expression during reperfusion of ischemic feline myocardium. *Circulation* 1993; 88:I2372.
24. Kubes P, Suzuki M, Granger DN. Nitric oxide—An endogenous modulator of leukocyte adhesion. *Proc Natl Acad Sci USA* 1991;88:4651–4655.
25. Dzau VJ, Gibbons GH. Introduction-Vascular remodeling: Mechanisms and implications. *J Cardiovasc Pharmacol* 1993;21(Suppl 1):S1–S5.
26. Stuehr DJ, Cho HJ, Kwon NS, Weise MF, Nathan CF. Purification and characterization of the cytokine-induced macrophage nitric oxide synthase—An FAD-containing and FMN-containing flavoprotein. *Proc Natl Acad Sci USA* 1991;88:7773–7777.
27. Nathan CF, Hibbs JB. Role of nitric oxide synthesis in macrophage antimicrobial activity. *Curr Opin Immunol* 1991; 3:65–70.
28. Moncada S, Palmer RMJ, Higgs EA. Nitric oxide—physiology, pathophysiology and pharmacology. *Pharmacol Rev* 1991; 43:109–142.
29. Pentros A, Bennett D, Vallance P. Effect of nitric oxide synthase inhibitors on hypotension in patients with septic shock. *Lancet* 1991;338:1557–1558.
30. Kilbourn RG, Jubran A, Gross SS, et al. Reversal of endotoxin-mediated shock by NG-methyl-L-arginine, an inhibitor of nitric oxide synthesis *Biochem Biophys Res Commun* 1990; 172:1132–1138.
31. Freudenberg H, Lichtlen PR. The normal wall segment in coronary stenoses—a postmortal study. *Z Kardiol* 1981;70:863–869.
32. Rubanyi GM. Reversal of hypercholesterolemia-induced endothelial dysfunction by L-arginine. *Circulation* 1991; 83:1118–1120.
33. Cooke J, Andon N, Girerd X, Hirsch A, Creager M. Arginine restores cholinergic relaxation of hypercholesterolemic rabbit thoracic aorta. *Circulation* 1991;83:1057–1062.
34. Drexler H, Zeiher AM, Meinzer K, Just H. Correction of endothelial dysfunction in coronary microcirculation of hypercholesterolaemic patients by L-arginine. *Lancet* 1991;338: 1546–1550.
35. Bassenge E, Stewart DJ. Effects of nitrates in various vascular sections and regions. *Z Kardiol* 1986;75:1–7.
36. Kurz MA, Lamping KG, Bates JN, Eastham CL, Marcus ML, Harrison DG. Mechanisms responsible for the heterogeneous coronary microvascular response to nitroglycerin. *Circ Res* 1991;68:847–855.
37. Bassenge E, Stewart DJ. Interdependence of pharmacologically-induced and endothelium-mediated coronary vasodilation in antianginal therapy. *Cardiovasc Drugs Ther* 1988;2:27–34.

38. Busse R, Pohl U, Mülsch A, Bassenge E. Modulation of the vasodilator action of SIN-1 by the endothelium. *J Cardiovasc Pharmacol* 1989;14:S81–S85.
39. Moncada S, Rees DD, Schulz R, Palmer RMJ. Development and mechanism of a supersensitivity to nitrovasodilators after inhibition of vascular nitric oxide synthesis in vivo. *Proc Natl Acad Sci* 1991;88:2166–2170.
40. Rafflenbeul W, Bassenge E, Lichtlen P. Competition between endothelium-dependent and nitroglycerin-induced coronary vasodilation (Konkurrenz zwischen endothelabhängiger und Nitroglycerin-induzierter koronarer Vasodilation) *Z Kardiol* 1989;78:45–57.
41. Jackson WF, Busse R. Elevated guanosine 3′-5′-cyclic monophosphate mediates the depression of nitrovasodilator reactivity in endothelium-intact blood vessels. *Naunyn Schmiedebergs Arch Pharmacol* 1991;344:345–350.
42. Bassenge E, Zanzinger J. Nitrates in different vascular beds, nitrate tolerance, and interactions with endothelial function. *Am J Cardiol* 1992;70:23b–29b.
43. Bassenge E, Mülsch A. Anti-ischemic actions molsidomine by venous and large coronary dilatation in combination with antiplatelet effects. *J Cardiovasc Pharmacol* 1989;14:S23–S28.
44. Bassenge E, Zanzinger J. Effectiveness of an NO-releasing pirsidomine derivative on coronary conductance during long-term administration. *J Cardiovasc Pharmacol* 1993;22(Suppl 7):S22–S26.
45. Feelisch M, Schönafinger K, Noack E. Thiol-mediated generation of nitric oxide accounts for the vasodilator action of furoxans. *Biochemical Pharmacology* 1992;44,No.6:1149–1157.
46. Stewart DJ, Elsner D, Sommer O, Holtz J, Bassenge E. Altered spectrum of nitroglycerin action in long term treatment: Nitroglycerin-specific venous tolerance with maintenance of arterial vasodepressor potency. *Circulation* 1986;74:573–582.
47. Rudolph W, Dirschinger J, Kraus F, Reiniger G, Hall D. Nitrate therapy in patients with coronary artery disease, preparation and doses with and without development of tolerance. *Z Kardiol* 1990;79(Suppl):III57–65.
48. Elkayam U. Tolerance to organic nitrates: Evidence, mechanisms, clinical relevance, and strategies for prevention. *Ann Intern Med* 1991;114:667–677.
49. Fung HL. Solving the mystery of nitrate tolerance. A new scent on the trail? *Circulation* 1993;88:322–324.
50. Parker JD, Farrell B, Fenton T, Cohanim M, Parker JO. Counter-regulatory responses to continuous and intermittent therapy with nitroglycerin. *Circulation* 1991;84:2336–2345.
51. Meszaros L, Bak J, Chu A. Cyclic ADP ribose as an endogenous regulator of the non-skeletal type ryanodine receptor Ca^{2+} channel. *Nature* 1993;364:76–79.
52. Watanabe H, Kakihana M, Ohtsuka S, Enomoto T, Yasui K, Sugishita Y. Platelet cyclic GMP: A potentially useful indicator to evaluate the effects of nitroglycerin and nitrate tolerance. *Circulation* 1993;88:29–36.
53. Bassenge E, Busse R, Pohl U. Freisetzung von EDRF aus Arterien: Ein neues antiaggregatorisches Prinzip. *Cor Vas* 1988;2:113–117.
54. Waldman SA, Rapoport RM, Ginsburg R, Murad F. Desensitization to nitroglycerin in vascular smooth muscle from rat and human. *Biochem Pharmacol* 1986;35:3525–3531.
55. Kowaluk EA, Seth P, Fung H-L. Metabolic activation of sodium nitroprusside to nitric oxide in vascular smooth muscle. *J Pharmacol Exp Ther* 1992;262:916–922.

56. Fishman AP. Endothelium: A distributed organ with diverse capabilities. *Ann NY Acad Sci* 1982;401:1–8.
57. Jeserich M, Munzel T, Just H, Drexler H. Reduced plasma L-arginine in hypercholesterolaemia. *Lancet* 1992;338:561.
58. Minor RL, Myers PR, Guerra R, Bates JN, Harrison DG. Diet-induced atherosclerosis increases the release of nitrogen oxides from rabbit aorta. *J Clin Invest* 1990;86:2109–2116.
59. Mügge A, Elwell JH, Peterson TE, Hofmeyer TG, Heistad DD, Harrison DG. Chronic treatment with polyethyleneglycolated superoxide dismutase partially restores endothelium-dependent vascular relaxations in cholesterol-fed rabbits. *Circ Res* 1991;69:1293–1300.
60. Holtz J, Giesler M, Bassenge E. Two dilatory mechanisms of anti-anginal drugs on epicardial coronary arteries in vivo: Indirect, flow-dependent, endothelium-mediated dilation and direct smooth muscle relaxation. *Z Kardiol* 1983;72:98–106.
61. Clozel JP, Veniant M, Hess P, Sprecher U. Effects of two angiotensin coverting enzyme inhibitors and hydralazine on coronary circulation in hypertensive rats. *Hypertension* 1991;18:II8–II14.
62. Lefer AM, Aoki N. Leukocyte-dependent and leukocyte-independent mechanisms of impairment of endothelium-mediated vasodilation. *Blood Vessels* 1990;27:162–168.
63. Garg UC, Hassid A. Nitric oxide-generating vasodilators and 8-bromo-cyclic guanosine monophosphate inhibit mitogenesis and proliferation of cultured rat vascular smooth muscle cells. *J Clin Invest* 1989;83:1774–1777.
64. von der Leyen H, Gibbons GH, Morishita R, Lewis NP, Zhang Y, Cooke JP, Dzau VJ. In vivo gene transfer to prevent neointima hyperplasia after vascular injury: Effect of overexpression of constitutive nitric oxide synthase (abstract). *FASEB J* 1994;8(5):A802.

Mechanisms of Action of Nitrates

Kristina E. Torfgård and Johan Ahlner

Summary. **Glyceryl trinitrate, isosorbide dinitrate, and isosorbide-5-mononitrate are organic nitrate esters commonly used in the treatment of angina pectoris, myocardial infarction, and congestive heart failure. Organic nitrate esters have a direct relaxant effect on vascular smooth muscles, and the dilation of coronary vessels improves oxygen supply to the myocardium. The dilation of peripheral veins, and in higher doses peripheral arteries, reduces preload and afterload, and thereby lowers myocardial oxygen consumption. Inhibition of platelet aggregation is another effect that is probably of therapeutic value. Effects on the central nervous system and the myocardium have been shown but not scrutinized for therapeutic importance. Both the relaxing effect on vascular smooth muscle and the effect on platelets are considered to be due to a stimulation of soluble guanylate cyclase by nitric oxide derived from the organic nitrate ester molecule through metabolization catalyzed by enzymes such as glutathione S-transferase, cytochrome P-450, and possibly esterases. The cyclic GMP produced by the guanylate cyclase acts via cGMP-dependent protein kinase. Ultimately, through various processes, the protein kinase lowers intracellular calcium; an increased uptake to and a decreased release from intracellular stores seem to be particularly important.**

Key Words. **organic nitrate esters, glyceryl trinitrate, nitroglycerin, isosorbide dinitrate, isosorbide-5-mononitrate, cyclic GMP**

*N*itroglycerin was introduced as a therapeutic agent for the treatment of angina pectoris more than a century ago [1], and since then several other nitro compounds with similar chemical properties have been introduced, for example, isosorbide dinitrate (ISDN), isosorbide-5-mononitrate (IS-5-MN), and pentaerythritol tetranitrate. An essential feature of all these compounds is a nitrate ester bond (R-O-NO_2). The nitrate ester group distinguishes the organic nitrate esters both from nitro compounds, which possess a carbogen-nitrogen bond (R-C-NO_2), and from compounds containing nitric oxide. The latter group includes sodium nitroprusside, and the nitro compounds include various explosives, such as trinitrotoluene and the fuel additive nitromethane. The terms *nitrate* vasodilators or nitrovasodilators have often been used to identify vasodilator compounds belonging to any of these groups. To clearly stress the important chemical and functional differences between nitrate esters and nitro- and nitric-oxide-containing drugs, we use the term *organic nitrate esters* exclusively when referring to drugs containing the -O-NO_2 moiety. This review will focus on organic nitrate esters commonly used in clinical practice, that is, glyceryl trinitrate (GTN), isosorbide dinitrate, and isosorbide-5-mononitrate, and will discuss the cellular mechanisms of the compounds in some detail. Tolerance development will not be dealt with in this paper.

All organic nitrate esters share similar pharmacologic properties, including a basic mechanism of action that exhibits only minor differences between the compounds; the functional importance of these slight differences is not clear. Differences in the chemical structure have consequences for the metabolism of the various organic nitrate esters: In order to exert their pharmacologic activities the compounds must be enzymatically metabolized to liberate nitric oxide (NO). This is in contrast to, for example, sodium nitroprusside (SNP).

It should also be mentioned that the metabolites of both GTN and ISDN are pharmacologically active and induce vascular smooth muscle relaxation but are less potent than their parent compounds [2,3]. This has long been appreciated as far as ISDN is concerned, and its main metabolite, IS-5-MN, has been introduced as an antianginal drug in its own right. Recent studies have shown that at least the dinitrate metabolites of GTN are active in humans and contribute to the vasodilating effect [4]. The structural formulae of the organic nitrate esters and their metabolites are shown in Figure 1.

Different Sites of Action

Clinically, organic nitrate esters are used as a remedy for angina pectoris, myocardial infarction, and congestive heart failure. The principal mechanism behind the therapeutic effect is the relaxation of vascular smooth muscle, which leads to vasodilation and hence unloading of the heart; however, other mechanisms might be of importance as well.

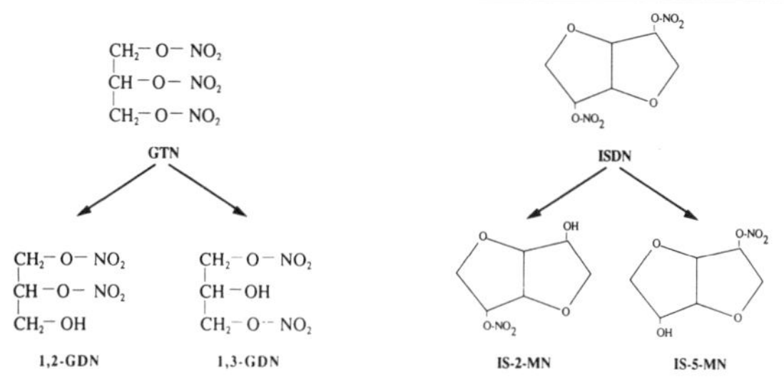

Fig. 1. *Chemical structures of some selected organic nitrate esters. Glyceryl trinitrate (GTN) and its dinitrate metabolites 1,2-glyceryl dinitrate (1,2-GDN) and 1,3-glyceryl dinitrate (1,3-GDN); isosorbide dinitrate (ISDN) and its mononitrate metabolites, isosorbide-2-mononitrate (IS-2-MN) and isosorbide-5-mononitrate (IS-5-MN).*

Blood vessels

Organic nitrate esters have a relaxant effect on all types of vessels, but the degree of the effect varies considerably between vessels from different anatomical regions and different species. In general, coronary arteries have been found to be more sensitive to glyceryl trinitrate than peripheral arteries [5–7]. Venous vessels are most sensitive to organic nitrate esters; this has been demonstrated both in vitro [8–11] and in vivo [12–14]. Toyoda and coworkers [10] compared the relaxation of rabbit femoral arteries and rabbit femoral veins and found that the latter were more sensitive and that the potency order between the three commonly used organic nitrate esters were GTN > ISDN > IS-5-MN. In other studies, a 10-fold differences in threshold concentration had been found between arterial and venous dilation, and IC_{50} values differed 10- to 100-fold [9,11]. These differences in sensitivity to organic nitrate esters might be explained by the finding that GTN, ISDN, and IS-5-MN inhibit Ca^{2+} release from Ca^{2+} stores more effectively in venous than in arterial preparations [15]. It was recently shown that GTN evokes a greater cGMP increase in the human internal mammary artery than in the saphenous vein and that comparatively less cGMP per milligram of protein is required for the maximal relaxation of the vein [16]. This is in accordance with the findings that veins are more sensitive to organic nitrate esters than arteries.

It has been proposed that nitrate concentrations should be higher in the tissues of the venous capacitance vessels than in arterial tissue. In support of this a bolus injection of GTN to rats resulted in a higher concentration in tissue from vena cava than from aorta [17]. However, comparing isolated bovine pulmonary artery and vein, Kawamoto and coworkers

[18] found that the concentration of glyceryl trinitrate was less and relaxation greater in the vein. In a study on patients undergoing coronary bypass surgery and receiving an intravenous infusion of GTN, we found that in general the concentration of GTN seemed to be higher in the internal mammary artery than in tissue taken from the vena saphena magma [19]. Several authors have reported that GTN-induced relaxation is augmented after removal of the endothelium [20–22], and it has been suggested that such "supersensitivity" after endothelial removal might be due to upregulation of soluble guanylate cyclase [22].

It should be kept in mind that in the clinical setting there are huge interindividual variations in the vasodilating effects of nitrates, particularly GTN. In a study by Eichler and coworkers [23], the GTN infusion rate that caused a 50% venodilation ranged from 0.2 to 22.4 ng/min, representing a 100-fold variation in the response of the effector organ. The venodilating effect of nitrates leads to a reduction of the left ventricular end-diastolic pressure and end-diastolic volume, and thus an unloading of the heart and a reduction of myocardial oxygen consumption. The decrease of the afterload, for which the arterial dilation is of importance, also unloads the heart and lowers oxygen consumption. Organic nitrate esters dilate coronary arteries, which leads to an increase in the oxygen supply to the myocardium. Not only are normal coronary arteries [24,25] dilated, but also collaterals [24] and even stenotic segments [25,26]. It has been demonstrated that GTN preferentially dilates larger (>200 µm in diameter) coronary microvessels [27]. The minimal effect on coronary resistance vessels (<100 µm in diameter) probably prevents the development of coronary steal, which is a problem with vasodilators (e.g., dipyridamole) acting on the smallest coronary microvessels. The lack of effect of GTN on smaller vessels might be due to reduced bioconversion of GTN [28].

Platelets

Several in vitro studies have shown that millimolar concentrations of GTN can inhibit platelet aggregation [29–31]. Very recently Chirkov and coworkers [32] observed that when nitroglycerin, even in vitro in concentrations >10 nM, was administered *after* platelet aggregation was induced, the aggregation was reversed. The maximal response to GTN was seen 0.5 minutes after the onset of aggregation. It may be that the responsiveness of platelet guanylate cyclase to nitric oxide is increased during aggregation, as proposed earlier by the same researchers

[33]. De Caterina and coworkers [34] showed a marked decrease in platelet aggregation in patients after an infusion of ISDN and, more recently [35], after infusions of IS-5-MN and IS-2-MN. During the past few years there have been reports of the beneficial effect of therapeutic doses of GTN on platelet function in humans [36,37]. Reduction of platelet deposition in the common carotid artery after endothelial injury has been documented in an animal model [38].

Myocardium

It has been suggested that organic nitrate esters might have a direct relaxant effect on cardiac muscle and thus a decreased inotropic effect [39,40]. There are, however, older reports that have shown GTN produces a positive inotropic effect on isolated feline papillary muscles [41] and on isolated blood-perfused canine papillary muscles [42]. Recently it was found that high concentrations (0.1 mM) of IS-2-MN inhibited the contractility of guinea-pig isolated heart muscle, whereas this influence was not exerted by IS-5-MN [43]. Organic nitrate esters seem to have a direct effect on myocardial metabolism, findings that may be in accordance with a negative inotropic effect. A number of studies on isolated rat atria have revealed beneficial effects of GTN. For example, under experimental conditions mimicking ischemia, GTN tended to normalize metabolism [44] and GTN has also been found to reduce oxygen consumption in isolated rat ventricular muscle, even in concentrations as low as 1 nM [45,46]. It must be stressed that it is not yet known to what extent a direct effect on myocardial metabolism contributes to the therapeutic effect of organic nitrate esters in ischemic heart disease.

Central nervous system

The most common side effect of organic nitrate esters is headache. This condition is supposed to be due to dilation of cerebral vessels, and nitroglycerin has been shown to dilate the temporal artery [47]. We have observed uptake of GTN by the brain and an increase in cGMP in brain tissue [48], and recently it was found that brain homogenates are able to metabolize GTN [49]. As early as 1960 it was proposed that the inhibition of reflex constriction of coronary vessels was mediated by increases in norepinephrine (NE) in the central nervous system (CNS) [50–51]. There are indications that GTN increases the release and synthesis of NE in brain tissues [52] and that cardiovascular effects of GTN involve a central as well as a peripheral component. Injections of GTN into the anterior hypothalamic preoptic area in rats was found to produce a dose-dependent decrease in mean arterial pressure (MAP) and heart rate [53], and it was proposed that GTN-induced modulation of noradrenergic transmission stimulates alpha$_2$-adrenoceptors in the CNS. It seems possible that both therapeutic and adverse effects are to some extent mediated via the CNS.

Other possible effects of therapeutic interest

During the past few years, intense research on the different effects of nitric oxide (NO) has disclosed that this simple molecule has a number of effects that might be of (therapeutic) interest when considering the NO-producing organic nitrate esters. It has been shown that NO inhibits not only platelet adhesion [54] but also leukocyte adhesion [55], endothelin generation [56], and smooth muscle cell proliferation [57]. NO seems to play a protective role in the oxidation of low-density lipoproteins [58]. Furthermore, NO reacts with superoxide anion; this reaction might represent detoxification of either molecule, that is, a scavenger function for NO. The importance of these findings for a possible antiatherosclerotic action of organic nitrate esters is currently unknown. There are indications that NO-cGMP has a direct effect on nociceptor sensitivity and consequently a peripheral analgesic effect that directly blocks hyperalgesia [59]. It was recently shown that topically applied GTN reduces prostaglandin E_2-induced hyperalgesia in rat skin [60]. Speculatively, this action might be of some importance for the antianginal effect of organic nitrate esters.

Different Modes of Action?

The pharmacological and hence therapeutic effects of the different organic nitrate esters are considered to be the same. However, of possible pertinence when discussing modes of action is whether there is more than one mechanism of action at the cellular level and whether there are differences between the different organic nitrate esters.

Different cellular mechanisms proposed

Several different mechanisms have been proposed to explain the relaxing effect of organic nitrate esters on vascular smooth muscle. On theory implies that there is a specific nitrate receptor, containing a sulfhydryl group (-SH group), in the smooth muscle cell membrane [61]; there are, however, no publications confirming the existence of such a receptor.

Another theory proposing that GTN inhibits thromboxane A_2 synthesis and stimulates prostaglandin synthesis was presented in the 1980s [62,63], but several workers have been unable to show that prostaglandins play a major role in mediating the vascular effect of organic nitrate esters [64–67]. Based on experimental findings, Rösen and coworkers [11] suggested that prostaglandins are of importance in mediating the GTN-induced relaxation represented by the high-affinity component of the concentration-response curve, that is, induced by low GTN concentrations. Such prostaglandin involvement in either the low- or the high-affinity component was later refuted by Malta [68]. A third theory about the smooth muscle relaxing effects of organic nitrate esters suggests the involvement of potassium channel activation. Recently Hamaguchi and coworkers showed that the relaxation of bovine tracheal smooth muscle by GTN and SNP involved charybdotoxin-sensitive (ChTx) K^+ channels [69]. These authors also observed that GTN, but not SNP, produced relaxation through an additional mechanism not sensitive to ChTx [69].

Some reports have indicated that cGMP-dependent processes might regulate Ca^{2+}-activated K^+ channel activity in vascular smooth muscle cells [70,71]. Activation of K^+ channels leads to hyperpolarization of the cell membrane, which is another theory that has been put forward in its own right to explain the smooth muscle relaxant effect of organic nitrate esters. Hyperpolarization of the cell membrane by GTN has been reported for isolated taenia coli of the guinea pig [72], the pig coronary artery [73], and the guinea pig portal vein [74] and uterine artery [75].

Today, it is generally agreed that the intracellular second-messenger cGMP mediates vascular smooth muscle relaxation elicited by the organic nitrate esters. The cyclic GMP system will be discussed in a separate section of this review, but it should be pointed out here that, for example, activation of potassium channels might be a part of, or an action in parallel with, a mechanism of action occurring via the cyclic GMP system.

Differences between different compounds?

Do the organic nitrate esters differ in effects and/or mechanisms of action? There are some reports indicating that arteries and veins vary in sensitivity to the different compounds. Stiefel and Kreye [76] found that ISDN and IS-5-MN were much more venoselective than GTN when comparing isolated renal arteries and veins from the rabbit, and similar results were reported in a study on rabbit femoral artery and vein [10]. There is only one limited clinical study in patients with chronic congestive heart failure that indicates a possible difference in venoselectivity between GTN and IS-5-MN [77], whereas there are several studies reporting no difference in this respect between the different organic nitrate esters [78–79].

Of some interest in this context is the possibility of enantioselectivity regarding the vascular effects of organic nitrate esters. The two stereoisometric forms of isosorbide dinitrate have been observed to exhibit a 10-fold difference in potency with regard to relaxation and cGMP elevation in isolated rat aorta: The d-isomer was most potent [80]. The two stereoisomers were equipotent in their ability to activate soluble guanylate cyclase, and it was suggested that the site of enantioselectivity was not the guanylate cyclase per se [81]. In a recent study we did not observe any difference in potency between two enantiomers of butanol-1,2,3,4-tetranitrate (erythritol tetranitrate and racemic thretiol tetranitrate) regarding the relaxation of bovine mesenteric activity [82]. This might indicate that enantioselectivity varies between different organic nitrate esters.

Several authors have found that GTN induces a biphasic pattern in vascular smooth muscle relaxation [10,11,20,66,82–85], which suggests the existence of two partially different mechanisms of action. In the literature, there is no evidence of a similar dual mechanism of action for ISDN or IS-5-MN-induced relaxation of vascular smooth muscle. Toyoda and coworkers [10] compared the effects of GTN, ISDN and IS-5-MN on femoral arteries and femoral veins from rabbits and observed the biphasic relaxation pattern only when the arteries were exposed to GTN. This is in accordance with studies on bovine mesenteric arteries performed by our research group [3,86]. Recently, we compared GTN to a number of newly synthesized organic nitrate ester analogues and, again, only GTN showed a biphasic relaxation pattern [82]. In contrast to other organic nitrate esters, the relaxation curve induced by GTN seems to consist of a high-affinity and a low-affinity component. This phenomenon will be further discussed in a later section in this review.

Metabolism

Several studies have suggested that organic nitrate esters act as prodrugs to release a pharmacologically active intermediate. The formation of this intermediate seems to depend on biotransformation or metabolism of the organic nitrate ester in or near the target tissue. It is now generally accepted that the relaxant action of organic nitrate esters on vascular smooth muscle occurs through NO and the cyclic GMP system, although the exact mechanism of the relaxation is not yet fully understood. Figure 2 presents a schematic drawing of the proposed mechanisms of the relaxant effect of organic nitrate esters. As indicated in the figure, NO is believed to be formed intracellularly [87], and to some extent, according to more recent studies, metabolic activation of GTN also occurs extracellularly [88], as well as in the cell membrane of the vascular smooth muscle cells [89].

The vasodilating effect of organic nitrate esters is coupled to the metabolism of the drug in the vascular tissue [18,90]. Furthermore, it has been shown that organic nitrate esters are metabolized in blood [91–94] as well as in homogenates of tissue samples from liver [93,95], lung and skeletal muscle [93,95], and intestine [93,96]; it was also recently confirmed that brain homogenates are able to metabolize GTN [49]. It is possible that the metabolism of organic nitrate esters to NO in tissues or structures adjacent to the vascular smooth muscle cell may be of importance for the vasodilating effect [19].

The vasodilating effect of GTN is less on small than on large microvessels [28], and this had been suggested to be due to a lower capacity for biotransformation of GTN to NO. Recent experiments have shown that the thiol-donating compound L-cysteine markedly increased the dilating effect of GTN in coronary microvessels <100 μM in diameter [97]. Another important factor to consider when discussing organic nitrate esters in regard to tissue metabo-

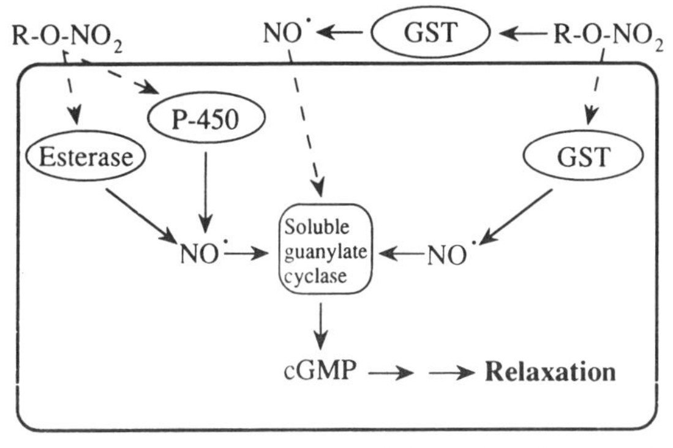

Fig. 2. *Schematic drawing of the proposed cellular mechanisms of action and metabolism of organic nitrate esters in the smooth muscle cell. R-O-NO_2 = organic nitrate ester; NO = nitric oxide; P-450 = cytochrome P-450; GST = glutathione S-transferase.*

lism and pharmacological effect is the actual activity of the metabolites. Although the principal metabolites of both ISDN and GTN are pharmacologically active and induce vascular smooth muscle relaxation, they are less potent than their respective parent compounds [2–4].

-SH groups and glutathione S-transferase

Chong and coworkers [98] have shown that -SH groups are required for the formation of active intermediates and that SH-containing compounds, such as cysteine or glutathione, are vital for the vasodilating action of organic nitrate esters. Furthermore, it has been reported that glutathione and glutathione S-transferase (GST) are involved in the metabolism of GTN by cultured smooth muscle cells [99], whereas it has also been shown in microsomes of the bovine coronary artery smooth muscle that it is not primarily GSTs that are responsible for the conversion of organic nitrate esters to NO [100]. It was recently shown that in vascular microsomes organic nitrate esters are converted to NO by a common enzyme that is not GST related; the reactivity of this NO-generating enzyme toward GTN and ISDN was consistent with the relative pharmacologic potencies reported for these compounds [101]. This membrane-bound enzyme responsible for generating NO from GTN in vascular smooth muscle cells has been characterized and requires a free thiol group for enzyme activity [102].

Several studies have, however, shown that S-nitrosothiols are produced from organic nitrate esters [61,87,94], which in turn stimulate guanylate cyclase to form cyclic GMP. Despite some results indicating otherwise, the formation of NO or S-nitrosothiols seems to depend on biotransformation of the organic nitrate esters in or near the target tissue. The involvement of -SH groups in the mechanism of action of organic nitrate esters is shown in Figure 2. This widely accepted theory has, however, been questioned by Sakanashi and coworkers, who showed that even when arterial glutathione content was almost completely depleted, the relaxant response to GTN was entirely unaffected [103]. In addition, studies have shown that GSTs present in vascular tissue do not contribute significantly to the bioconversion of NO [28,104]. Kurz and coworkers suggest that GSTs may represent a competitive pathway for the metabolism of GTN by converting GTN to nonvasoactive NO_2^- and GDN [28].

Recent experiments on liver homogenates have shown that both the cytosolic and microsomal fraction metabolize GTN, although the formation of GDNs seemed to be governed by different enzymatic processes in the two: 1,2-GDN was preferentially formed in the cytosolic fraction, whereas 1,3-GDN was the predominant product in the microsomal fraction. Furthermore, the two GTN metabolites differed in sensitivity to inhibitors of GST. Based on these findings, it was concluded that the formation of 1,2-GDN and 1,3-GDN must be catalyzed by different isoenzymes of glutathione S-transferase [105]. GST had been isolated from human erythrocytes [106], and heart and aorta [107]. Recently the effects of sulfobromophtalein (SBP) and ethacrynic acid (ECA), both inhibitors of GST and GTN-induced vasorelaxation, were investigated in rabbit aorta strips. ECA inhibited GTN metabolism significantly more than SBP did, and in aortic strips pre- contracted with phenylephrine, ECA inhibited GTN relaxation, whereas SBP did not alter the GTN activity. These results suggest that the mechanisms of GTN activation may involve a GST isoenzyme that is highly active towards ECA [108], which is in contrast to an earlier report indicating that GTN-induced relaxation can be antagonized by SBP [109].

However, Nigham and coworkers showed recently that a number of GSTs inhibitors show different results depending on the nature of the contractile agent used [110]. It was also shown that the GSTs inhibitor Basilen Blue inhibited GTN biotransformation in tissues precontracted with K^+.

Cytochrome P-450 and esterases

Recent studies have demonstrated that hepatic microsomal reductive denitration of GTN is catalyzed by cytochrome P-450 (P-450) [111,112], and P-450 has also been identified in extrahepatic tissues, such as the intima of the aorta [113]. Experiments with the P-450 inhibitors SKF-525A and metyrapone and general hemeprotein inhibitors (CO and NO) failed to inhibit GTN-induced relaxation of rabbit aortic preparations or GTN biotransformation to GDN [114]. The P-450 substrate, 7-ethoxyresorufin (7-ER) has been reported to inhibit vascular P-450, and Bennett and coworkers showed recently that 7-ER inhibited the GTN-induced relaxation and cGMP accumulation in isolated rat aortic strips, and that 7-ER selectively reduced the 1,2-GDN-formation but not the formation of 1,3-GDN [115]. Experiments on kidney epithelial cells and lung fibroblasts have shown that the GTN-induced cyclic GMP stimulation in intact cells is mediated by a P-450 enzyme system rather than by GST or free thiols; however, cultured porcine aortic endothelial cells did not respond to GTN with an increase in cyclic GMP [116,117]. Furthermore, Bennett and coworkers [118] have shown a thiol-independent activation of aortic guanylate cyclase by GTN in the presence of hepatic microsomes and have also presented strong evidence that this effect is mediated by P-450. These authors also found that both type I and type II P-450 inhibitors are able to block NO formation during microsomal biotransformation of GTN. In contrast, P-450 inhibitors had no effect on GTN-induced relaxation, whereas P-450 inducers increased the sensitivity of rat aorta to GTN. Bennett and coworkers also showed that the sensitivity of isolated rat aorta to GTN-induced relaxation increased under low oxygen conditions. This increased sensitivity was associated with an increase in the vascular biotransformation of GTN, suggesting that vascular biotransformation of GTN is at least partially mediated by hemoproteins. The metabolism of GTN and the formation of NO seems to depend on a P-450-dependent biotransformation of the organic nitrate esters in or near the target tissue. The proposed involvement of P-450 in the mechanisms of action of organic nitrate esters is shown in Figure 2.

It has also been put forward that esterases may be involved in the metabolism of organic nitrate esters. However, thus far studies have been performed only to determine what effects these enzymes have on the distribution of GTN [19]. Esterases have been found in plasma, erythrocytes, liver, and gut [119], and may therefore be involved in the biotransformation/metabolism of GTN and the formation of NO in the blood. If esterases are present in vascular smooth muscle, it is also possible that GTN may be denitrated by these enzymes in the vessels themselves. Another possibility is that GTN is metabolized by esterases in tissues or structures adjacent to the vascular smooth muscle, and that this NO formation might also be of importance for vascular smooth muscle relaxation. The proposed involvement of esterases in the mechanisms of action of organic nitrate esters is shown in Figure 2.

It has been shown that lipophilicity is of importance for the vasodilating effect of organic nitrate esters [120], which suggests that the number of NO-containing groups is important not only for lipophilicity, but also for the relaxing effect of organic nitrate esters due to the availability to biotransformation/metabolism. In conclusion, several different enzymes may be involved in the metabolism of GTN and the formation of NO, but their mode of involvement is not known.

Cyclic GMP System

The vascular smooth muscle relaxation induced by organic nitrate esters is believed to be mediated via the intracellular second messenger cyclic GMP. The formation of cyclic GMP from guanosine 5′-triphosphate is catalyzed by the enzyme guanylate cyclase, of which there are at least three forms: a membrane-bound particulate guanylate cyclase, which can be activated by atrial natriuretic peptides (ANP) and various egg peptides; a form associated with the cytoskeleton can be activated by *E. coli* heat-stable entertotoxin and calcium; a soluble form, which is the major target for activation by organic nitrate esters.

Soluble guanylate cyclase

Soluble guanylate cyclase has been purified to apparent homogeneity from different sources, including human platelets [121]. The enzyme has been characterized as a heterodimer, the two subunits of which have apparent molecular weights of 82 and 70 kD [121–123]. Expression of enzymatic activity and stimulation by nitric oxide appear to require both subunits [124,125]. Purified guanylate cyclase contains heme [126,127], which is important for the activation of the enzyme by nitric oxide and nitric oxide-releasing agents, such as organic nitrate esters. In electron paramagnetic resonance studies it was demonstrated that the binding of nitric oxide to heme results in a structural change of the heme with a displacement of the iron atom from the plane of the protoporphyrin molecule [128,129]. In two subsequent studies, it was suggested that nitric oxide activates guanylate cyclase by conformationally changing the heme prosthetic group so that the Fe^{2+} atom protrudes out of the plane of the protoporphyrin molecule [130–131]. Thiol compounds and thiol-modifying agents have pronounced effects on guanylate cyclase activity, and this is of significance for the effects of organic nitrate esters. There are three possible interactions between thiols and guanylate cyclase. First, thiol groups located on the enzyme seem to be important for the regulation of the enzyme activity, probably through reversible formation of disulfide bridges between juxtaposed SH groups [132–133]. Second, thiol groups may facilitate the generation of reactive intermediates from organic nitrate esters and related substances [89,134–136]; these intermediates formed are either nitric oxide or S-nitrosothiols, and can function as the ultimate activator of guanylate cyclase, probably via an interaction with the heme moiety of the enzyme. Third, thiols may also directly facilitate the formation of nitrosyl-heme complexes by acting as reducing agents and thus maintaining the heme iron in the reduced (Fe^{2+}) state [128,137,138].

Cyclic GMP

The cyclic GMP formed by the soluble guanylate cyclase exerts its main action through activation of cGMP-dependent protein kinase (cG-Pk), at least in vascular smooth muscle and platelets. There are several cG-Pk isoenzymes with different chemical characteristics. Binding of cGMP to cG-Pk results in activation of the enzyme. Several different proteins that are phosphorylated by cG-Pk have been described, although the significance of these proteins in regulating cell functions remains unclear (see also subsequent sections in this review).

The action of cGMP is terminated when the 3′,5′-cyclic phosphoester bond is opened, a process that yields GMP, which is unable to activate cG-Pk; inactivation of cGMP is catalyzed by phosphodiesterases. Five distinct families of phosphodiesterases have been identified, one of which is cGMP specific [139]. Some selective inhibitors of the cGMP-specific family have been synthesized, for example, dipyridamole and zaprinast [140–141], and have proven to be valuable tools for elucidating the mechanisms of action of organic nitrate esters. Activation of the cyclic GMP system in the vascular smooth muscle cell is illustrated in Figure 3.

Cyclic GMP and vascular smooth muscle relaxation

It is generally accepted that cGMP mediates the vascular effects of organic nitrate esters and also other clinically important effects on different target organs, such as platelets. Sutherland and coworkers set up

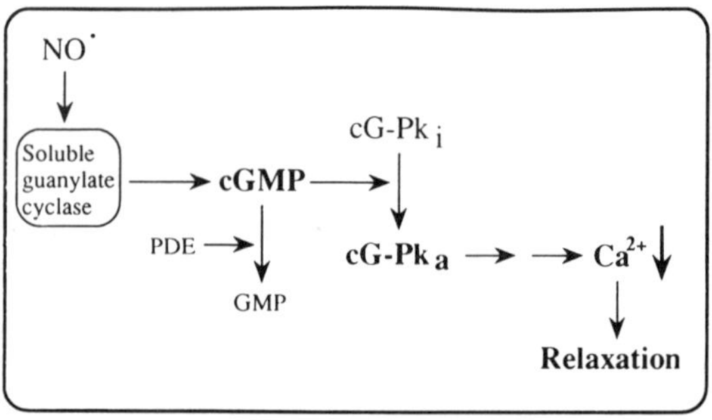

Fig. 3. *Schematic drawing of activation of the cGMP system in the smooth muscle cell. NO = nitric oxide; PDE = phosphodiesterase; cG-Pk = cGMP-dependent protein kinase; i = inactive; a = active.*

criteria for proving a causal relationship between an intracellular pathway and a response elicited by a defined stimulus [142]. Considering existing research results, the criteria are met by cGMP as the intracellular mediator of the vascular smooth muscle relaxing effect of organic nitrate esters. Several in vitro studies have demonstrated a close correlation between vascular smooth muscle relaxation and the increase in cGMP. This relationship has been established for various blood vessels from different species and for different organic nitrate esters, including those most commonly used, that is, GTN, ISDN, and IS-5-MN [46]. Also in preparations of various types of vessels and using different organic nitrate esters, it has been noted that relaxation is *preceded* by an increase in tissue cGMP level. Demonstrating such a temporal relationship in vivo is much more difficult, although it has been found that a cGMP increase precedes an increase in coronary blood flow in rats [143]. Potentiation of vascular smooth muscle relaxation and cyclic nucleotide levels by various inhibitors of cyclic nucleotide phosphodiesterases has been reported. Both zaprinast and dipyridamole, which are selective inhibitors of cGMP-specific phosphodiesterases, have been shown to potentiate the relaxation induced by organic nitrate esters in various isolated vascular smooth muscles [5,31,144–146]. It had also been shown that plasma-membrane-penetrating cGMP analogues, chiefly 8-bromo-cGMP, can induce relaxation in vascular smooth muscle preparations and that this relaxation is potentiated by phosphodiesterase inhibitors [144,147–150]. Only one of the classical criteria proposed by Sutherland and coworkers remains to be satisfied, namely, the one stating that, whether in intact tissue or broken cell preparations, enzymes involved in the formation of the signal molecule should respond to the same stimuli. Significant guanylate cyclase activation in a crude tissue homogenate or a crude soluble fraction can be obtained only at high concentrations (micromolar to millimolar range) of organic nitrate esters and in the presence of exogenously added thiol compounds [136,151,152].

Glyceryl trinitrate—biphasic relaxation pattern

Studies on isolated vascular preparations have indicated that when considering a concentration-effect relationship, GTN differs from other organic nitrate esters. The mechanisms behind the two different components of the biphasic concentration-effect curve for GTN are unclear. However, both components do seem to be mediated through an elevation in cGMP [20,85], since they are inhibited by substances that suppress cGMP generation, for example, methylene blue and 6-anilino-5,8-quinolinedione (LY83583) [85,86]. On the other hand, the mechanisms of action of the components differ in that inhibition of the high-affinity component is noncompetitive with regard to GTN, whereas the inhibition of the low-affinity component is competitive [85]. The two response-curve components are affected differently by tolerance development [11,86]. It has been shown that induction of "low-dose" tolerance in bovine mesenteric arteries (0.1 nM GTN, 2 hours) preferentially reduces the high-affinity component and has only a minor effect on the low-affinity component, whereas "high-dose" tolerance (100 μM GTN, 2 hours) reduces both the low- and high-affinity components [86]. These findings support the assumption that the different components are mediated, at least partly, through separate mechanisms.

In addition to the above-mentioned characteristics of the biphasic concentration-effect relationship for GTN, there are differences between the high-affinity and the low-affinity components regarding the sensitivity to pertussis toxin. Pertussis toxin is widely used as a pharmacological tool for establishing the involvement of certain types of guanine regulatory proteins in cellular signal transduction pathways [153,154]. ADP ribosylation of susceptible G proteins by pertussis toxin induces characteristic changes in the function of these proteins, resulting in disturbed signal transduction. Incubation of bovine mesenteric arteries with pertussis toxin significantly reduced the relaxant effect of GTN [155]; this inhibition was comparatively more pronounced for the high-affinity component. The effect of pertussis toxin was accompanied by an abolished cGMP response to GTN in the concentration range representing the high-affinity component, and the effect of pertussis toxin was antagonized by a polyclonal antibody directed against pertussis toxin [155].

Theoretically, the biphasic concentration-effect curve for GTN could be due to activation of different pools of soluble guanylate cyclase [20]. The activation of such pools would, at least to some extent, proceed via separate pathways, and a pertussis toxin-sensitive regulatory protein would be involved in activation of the pathway representing the high-affinity component. However, the effect of pertussis toxin may not be related to modification of a regulatory protein in the traditional manner. It has been shown that the site of ADP ribosylation of G proteins by pertussis toxin is a conserved cysteine residue located four amino acids from the COOH terminus of the αsubunit of the proteins [153], and thiol compounds may also function as ADP ribosylation substrates for pertussis toxin in vitro [156]. The effect of pertussis toxin on relaxation and cGMP production might be due to inactivation of critical thiol groups necessary for the activation of soluble guanylate cyclase. These thiol groups could be located either on the guanylate cyclase itself or on proteins/peptides involved in the production of guanylate cyclase-activating intermediates (e.g., S-nitrosothiols) from GTN. Another possible explanation for the biphasic concentration-effect curve produced by GTN is that different isoenzymes of guanylate cyclase may be activated. Heterogeneity has been reported for soluble guanylate cyclase [157,158], and it is possible that the different forms are differentially regulated. Moreover, nitric oxide activation of washed and partially purified particulate forms of guanylate cyclase has been demonstrated [159–161]. It is also possible that the particulate form of guanylate cyclase associated with the cytoskeleton may be activated by organic nitrate esters and thus contribute to the biphasic response. Little is known about this enzyme in tissues other than intestinal mucosa or about its possible activation by agents other than *E. coli* heat-stable enterotoxin. Interestingly, the signal transduction between the receptor for *E. coli* heat-stable enterotoxin and the effects mediated by this cytoskeleton-associated guanylate cyclase appears to be sensitive to pertussis toxin [162]. The idea that the different components of the concentration-effect curve for GTN may be explained by activation of different pools or isoenzymes of guanylate cyclase may also imply that cellular compartmentalization exists with regard to cGMP formed by these different mechanisms. Such compartmentalization would in turn lead to activation of distinct cellular processes (e.g., reduction in cytosolic free calcium concentration, phosphorylation of contractile proteins, hyperpolarization) and ultimately lead to vascular smooth muscle relaxation. Even if the relevance of the mechanism behind the biphasic relaxation pattern for GTN is totally unknown, it should be mentioned that compartmentalization of cyclic GMP-dependent protein kinase in vascular smooth muscle cells has recently been proposed [163].

Cyclic GMP as mediator of effects on platelet and myocardium

Cyclic GMP seems to mediate inhibition of platelet aggregation exerted by organic nitrate esters, since such inhibition is correlated with an increase in cGMP, and both the antiaggregatory action and the elevation in cGMP are potentiated by dipyridamole (a cGMP-phosphodiesterase inhibitor) and N-acetylcysteine [30,31,164]. Furthermore, it has been suggested that platelet adhesion and the platelet release reaction may be regulated by cGMP [165–169]. It was recently shown that pretreatment of platelets with the cGMP analogue 8-bromo-cGMP significantly reduces platelet adhesion to endothelium [170] and that incubation of platelets with exogenous cGMP results in inhibition of platelet secretion [171]. It has also been suggested that soluble guanylate cyclase of platelets is involved in the regulation of stimulated platelet PAI-1 release [172]. Platelets contain high concentrations of cGMP-dependent protein kinase, and the phosphorylation of a specific protein is stimulated by agents (e.g., SNP) that increase the cGMP level in platelets [173].

Direct effects of organic nitrate esters on myocardial cells have been demonstrated under certain conditions. The lactate content was reduced by 8-Br-cGMP and SNP in spontaneously beating ischemic rat atrial tissue in vitro [44,174]. GTN in the presence of cysteine was found to have a similar effect, whereas GTN alone was ineffective in reducing lactate levels in ischemic rat atrium and was also without effect on the cGMP level [175]. In addition, an increase in tissue energy charge and a reduction in nicotinamide adenine nucleotide (reduced form) levels have been demonstrated, indicating an overall improvement of metabolic parameters by cGMP during ischemia [44,176,177]; this has also been demonstrated in the Langendorf-perfused rat heart [178]. However, in a recent study in our laboratory, GTN, in the concentration range 1 nM to 1 mM, was found to increase lactate accumulation in rat ventricular muscle, under normoxic conditions and during hypoxia. In contrast, both SNP and ANP was found to reduce lactate production in ischemic left ventricle in vitro [46,179]. Moreover, SNP and ANP increased cGMP levels in ischemic myocardium, whereas GTN did not.

It has been shown that twitch contractions of ferret cardiac muscle are shortened by 8-bromo-cyclic GMP [180]. In addition, cGMP has been demonstrated to block the positive inotropic effect of α_1 stimulation in ferret preparations [181]; the mechanism in this case was a blockade of the increase in inositol phosphate (IP_3). Another possibility for a cGMP-mediated effect on myocardial contraction is a phosphorylation of cardiac troponin I by cyclic GMP-dependent protein kinase [182,183], with a consequent reduction in the affinity of troponin c for calcium [184]. Cyclic GMP has also been reported to reduce calcium influx in amphibian cardiac muscle [185]. Thus it seems clear that the effects exerted by organic nitrate esters on vasodilation and platelet aggregation are mediated by cGMP. The direct effect on the myocardium deserves further study, since possible therapeutic importance and the basal mechanisms involved are unclear. Research in this area will probably be intensified because it is now known that nitric oxide (the endothelium-derived relaxing factor) is produced by both the endocardial endothelium and vascular endothelium, and also seems to be able to influence contraction of underlying cardiac muscle [186].

Decreased intracellular Ca-concentration—The ultimate step?

Intracellular free calcium concentration plays a central role in the regulation of smooth muscle tension. The vascular smooth muscle relaxation and the inhibition of platelet aggregation and platelet adhesion induced by cGMP is due to a reduction in cytosolic calcium concentration. Cyclic GMP-dependent protein kinase is one of the major factors regulating cytosolic calcium levels [187].

Organic nitrate esters and other agents causing cGMP elevation have been shown to reduce the intracellular concentration of free calcium. This has been studied by indirect methods [188–190] and, primarily, more recently, by direct methods, with which cGMP, ANP, and SNP have been found to reduce the intracellular calcium concentration in both vascular smooth muscle cells and platelets [191–193]. GTN and ISDN have also been shown to inhibit receptor-induced increase in cytosolic calcium in human platelets [194]. A decrease in the concentration of free intracellular calcium can be achieved by the following: a reduction in calcium influx across the plasma membrane, an increased binding of calcium to intracellular stores, a reduced calcium release from these stores, and an increased outflow of calcium from the cell.

Grün and coworkers suggested that GTN has a calcium-antagonistic effect in vascular smooth muscle [195–196]. In studies using radiolabeled calcium, SNP was reported to reduce calcium influx and/or to increase calcium efflux in vascular smooth muscle preparations, and it was suggested that this is also a major mechanism of action for organinc nitrate esters [197–200]. A reduction in calcium influx has been reported in rabbit pulmonary artery as well, but only when millimolar concentrations of GTN were used [201]. In intact bovine mesenteric arteries in vitro, 1 μM GTN was found to decrease phenylephrine-stimulated calcium influx [202], whereas a lower concentration (0.1 nM) exerted a substantial relaxant effect without any effect on calcium influx. In other studies, GTN and SNP have not been found to have any effect on transmembrane calcium fluxes [203–204]. Thus, organic nitrate esters seem to exert some effect on calcium fluxes across the plasma membrane, but only in higher concentrations and in certain types of vascular smooth muscle.

Increased binding of calcium to intracellular stores in arterial smooth muscle is involved in the relaxant activity of organic nitrate vasodilators, as has been shown for SNP, GTN, and cGMP in various vascular smooth muscle preparations [200,202,205,206]. It was recently observed that cyclic GMP-dependent protein kinase activates sarcoplasmic reticulum Ca^{2+}-ATPase via phosphorylation of phospholamban [163]. This leads to a decrease in cytosolic free Ca-concentration due to increased calcium uptake by the sarcoplasmic reticulum.

We have shown that IP_3-stimulated calcium release from microsomes is significantly reduced by GTN, both at low (0.1 nM) and high (1 mM) concentrations [202]. This effect is transient and is mimicked by exogenously applied cGMP. Cyclic GMP and substances that stimulate cGMP synthesis, among them GTN, can reduce phosphatidylinositol turnover in isolated vascular smooth muscle and platelets [165,207–210]. A reduction in IP_3 production, which in turn reduces calcium release from intracellular stores, might be a mechanism by which cGMP and organic nitrate esters induce vascular smooth muscle relaxation.

Protein kinase C seems to participate in the regulation of smooth muscle tension and platelet function, probably through phosphorylation of various proteins [211–214]. Hence, inhibition of protein kinase C could facilitate relaxation of vascular smooth muscle and inhibition of platelet aggregation. In a previous study we were able to show that GTN inhibits the protein kinase C activation induced by contractile stimuli [210]. Such an effect might be due to decreased intracellular free calcium concentration and/or decreased formation of diacylglycerol, both of which are important activators of protein kinase C. Diacylglycerol and IP_3 are formed by phospholipase C, and it has been proposed that cGMP inhibits this enzyme, possibly through modification of a guanine regulatory protein [207,215]. Hirata and coworkers [215] suggested that this is achieved via cGMP kinase inhibition the GTPase activity of one or several G-protein(s), thereby reducing the activation of phospholipase C. However, this idea is not undisputed. Lincoln [216] found that neither pertussis toxin-sensitive nor pertussis toxin-insensitive G proteins were phosphorylated by cGMP-dependent protein kinase. Another way in which organic nitrate esters and cGMP can regulate intracellular calcium concentration is via stimulation of the sarcolemmal calcium-pumping ATPase; this has been shown in a number of studies [202,217,218]. A recent paper suggests that phosphorylation of a 240-kD membrane protein by cGMP protein kinase might be the mechanism behind this effect [219].

Even if it seems clear that organic nitrate esters act via cGMP to reduce intracellular calcium levels, the functional importance of this effect has been questioned. Simultaneous measurements of force and intracellular calcium concentrations in one study showed that GTN-induced relaxation was clearly not associated with a decrease in intracellular calcium. Relaxation occurred without any reduction of intracellular calcium [220]. However, the curve describing the relationship between intracellular free [Ca^{2+}] and force was shifted to the right, indicating an effect of GTN on the contractile proteins (presumably a decrease in myosin light chain phosphorylation) [220]. Studies of phosphorylation of endogenous proteins have indicated that cGMP reduces the phosphorylation of myosin light chain in vascular smooth muscle and platelets, and also increases the phosphorylation of a number of protein with unknown functions [191,221–224].

It was recently reported that nitrovasodilator-induced relaxation of histamine-stimulated swine arterial smooth muscle involves an uncoupling of stress from myosin phosphorylation in addition to a reduction of free intracellular calcium levels [225]. It is believed that the former mechanism is not dependent on an increase in cGMP. It should be mentioned that other actions of NO not linked to cGMP have also been demonstrated. NO (and isosorbide dinitrate) has been found to decrease cytosolic free calcium in a type of fibroblasts known to lack soluble guanylate cyclase activity, and neither cGMP nor 8-bromo-cGMP mimicked this effect when added to the same type of cells [226]. Several authors have been able to show that NO stimulates the ADP ribosylation of a protein present in human platelets and various other tissues [227]. However, the importance of these mechanisms for the therapeutic effects of organic nitrate esters is totally unknown. It seems reasonable to conclude that a reduction of intracellular free calcium is of importance for both the vascular smooth muscle relaxant effect and the inhibition of platelet aggregation evoked by organic nitrate esters, and that cGMP is an important mediator of these effects. This does not, however, rule out the possibility that other mechanisms might be involved as well.

Conclusions

Organic nitrate esters are still a mainstay in the treatment of angina pectoris, and they are also used in myocardial infarction and congestive heart failure. Organic nitrate esters have a direct relaxant effect on vascular smooth muscles, an effect that leads to dilation of coronary vessels and thereby improved oxygen supply to the myocardium. The dilation of peripheral veins, and in higher doses peripheral arteries, reduces preload and afterload, which in turn lowers myocardial oxygen consumption. Inhibition of platelet aggregation is another effect that is probably of therapeutic value. Both the vascular smooth muscle relaxing effect and the effect on platelets are due to stimulation of soluble guanylate cyclase by nitric oxide derived from the organic nitrate ester molecule through metabolism catalyzed by enzymes (e.g., glutathione S-transferase, cytochrome P-450, and possibly esterases). The cyclic GMP produced by the guanylate cyclase acts via cGMP-dependent protein kinase to lower levels of intracellular free calcium. In particular, two processes seem to be affected: The uptake of calcium to intracellular stores is increased and the release of calcium from intracellular stores is decreased.

Continuing research regarding nitric oxide and its mechanisms of action will provide increasing knowledge of the basal mechanisms behind the action of organic nitrate esters. Consequently, the mechanisms behind both tolerance development and the great interindividual differences in effect exhibited by organic nitrate esters will be better understood. Of great value in this respect would be the possibility of studying the importance of transcriptional and translational regulation of guanylate cyclase that would be provided by successful cloning of soluble guanylate cyclase. Despite the long tradition of clinical use of organic nitrates, a number of intriguing questions regarding the basal mechanisms of action of these compounds remain unanswered.

Acknowledgments

This paper was written with financial support from Östergötlands landsting, Lions forskningsfond, and from Konung Gustaf V:s 80 års fond.

References

1. Murrel W. Nitro-glycerine as a remedy for angina pectoris. *Lancet* 1879;1:80–81, 113–115, 151–152, 225–227.
2. Axelsson KL, Andersson RGG, Wikberg JES, et al. Correlation between vascular smooth muscle relaxation and increase in cyclic GMP induced by some organic nitro esters. *Acta Pharmacol Toxicol* 1981;49:270–276.
3. Torfgård K, Ahlner J, Axelsson KL, et al. Relaxation of bovine mesenteric arteries by glyceryl trinitrate and other stimulators of soluble guanylate cyclase. Evidence for different mechanisms of action. *Pharmacol Toxicol* 1990;67:216–221.
4. Gumbleton M, Benet LZ. Pharmacological activity of the dinitrate metabolites of nitroglycerin following their oral administration to healthy volunteers. *Br J Clin Pharmacol* 1991;31:221–213.
5. Shibata T, Ogawa K, Ito T, et al. Role of cyclic GMP of canine vascular smooth muscle relaxation by organic nitroesters. *Jpn Circ J* 1986;50:1091–1099.
6. Gharaibeh MN, Gross GJ. Comparative relaxant effects of nitroglycerin in isolated rings from various canine vascular beds. *Gen Pharmacol* 1984;15:217–221.
7. Miwa K, Toda N. The regional difference of relaxations induced by various vasodilators in isolated dog coronary and mesenteric arteries. *Jpn J Pharmacol* 1985;38:313–330.
8. Mackenzie JE, Parratt JR. Comparative effects of glyceryl trinitrate on venous and arterial smooth muscle in vitro: Relevance to antianginal activity. *Br J Pharmacol* 1977;29:155–160.
9. Nakajima H, Nosaka K. A comparison of the effects of diltiazem and nitroglycerin on the norepinephrine-induced contractions in the isolated femoral artery and vein. *Jpn J Pharmacol* 1983;33:1282–1285.
10. Toyoda J, Hisayama T, Takayanagi I. Nitro compounds (isosorbide dinitrate, 5-isosorbide mononitrate and glyceryl trinitrate) on the femoral vein and femoral artery. *Gen Pharmacol* 1986;17:89–91.
11. Rösen R, König E, Klaus W. Different sensitivities of arteries and veins to glyceryl trinitrate-induced relaxation and tolerance: An “in vitro” study on isolated vessels in rabbits. *Arch Int Pharmacodyn Ther* 1987;285:226–237.
12. Imhof PR, Ott B, Frankhauser P, et al. Differences in GTN dose-response in the venous and arterial beds. *Eur J Clin Pharmacol* 1980;18:455–460.
13. Miller RR, Vismara LA, Williams D, et al. Pharmacological mechanisms for left ventricular unloading in clinical congestive heart failure: Differential effects of nitroprusside, phentolamine and nitroglycerin on cardiac function and peripheral circulation. *Circ Res* 1976;39:127–133.
14. Gerson JL, Allen FB, Seitzer JL, et al. Arterial and venous dilation by nitroprusside and nitroglycerin—is there a difference? *Anesth Analg (Cleve)* 1982;61:256–260.
15. Toyoda J, Hisayama T, Takayanagi I. Effects of nitro compounds, isosorbide dinitrate, 5-isosobride mononitrate and glyceryl trinitrate on Ca-uptake into Ca-stores and Ca-release from Ca-stores in rabbit isolated femoral veins and femoral arteries. *Gen Pharmacol* 1987;18:95–97.
16. Tadjkarimi S, O'Neil GS, Luu TN, et al. Comparison of cyclic GMP in human internal mammary artery and saphenous vein: Implications for coronary artery bypass graft patency. *Cardiovasc Res* 1992;26:297–300.

17. Fung H-L, Sutton SC, Kamiya A. Blood vessel uptake and metabolism of organic nitrates in the rat. *J Pharmacol Exp Ther* 1984;228:334–341.
18. Kawamoto JH, Brien JF, Marks GS, et al. A comparative study of glyceryl trinitrate biotransformation and glyceryl trinitrate induced relaxation in bovine pulmonary artery and vein. *Can J Physiol Pharmacol* 1987;65:1146–1150.
19. Torfgård KE. Glyceryl Trinitrate Distribution and the "Remarkable Tolerance." Linköping University Medical Dissertations No. 358, Linköping, 1992, pp 1–68.
20. Ahlner J, Axelsson KL, Ekstam Ljusegren M, et al. Demonstration of a high affinity component of glyceryl trinitrate induced vasodilation in the bovine mesenteric artery. *J Cyclic Nucl Prot Phosphoryl Res* 1987;11:445–456.
21. Dinerman JL, Lawson DL, Metha JL. Interactions between nitroglycerin and endothelium in vascular smooth muscle relaxation. *Am J Physiol* 1991;260:H698–H701.
22. Moncada S, Rees DD, Schulz R, et al. Development and the mechanism of a specific supersensitivity to nitrovasodilators after inhibition of vascular nitric oxide synthesis in vivo. *Proc Natl Acad Sci USA* 1991;88:2166–2170.
23. Eichler H-G, Hiremath A, Katzir D, et al. Absence of age-related changes in venous responsiveness to nitroglycerin in vivo in humans. *Clin Pharmacol Ther* 1987;42:521–524.
24. Feldman RL, Pepine CJ, Curry RC, et al. Coronary arterial responses to graded doses of nitroglycerin. *Am J Cardiol* 1979;43:91–97.
25. Brown BG, Bolson E, Petersen RB. The mechanism of nitroglycerin action: Stenosis dilation as a major component of drug response. *Circulation* 1981;64:1089–1097.
26. Sievert H, Selzer G, Schneider W, et al. Coronary stenosis dilation by low dose intravenous nitroglycerin. *Eur Heart J* 1989;10:134–136.
27. Sellke FW, Myers PR, Bates JN, et al. Influence of vessel size on the sensitivity of porcine coronary microvessels to nitroglycerin. *Am J Physiol* 1990;258:H515–H520.
28. Kurz MA, Boyer TD, Whalen R, et al. Nitroglycerin metabolism in vascular tissue: Role of gluathione S-transferases and relationship between NO^- and NO_2^- formation. *Biochem J* 1993;292:545–550.
29. Schaffer AI, Alexander RW, Handin RI. Inhibition of platelet function by organic nitrate vasodilators. *Blood* 1980;55:649–654.
30. Loscalzo J. N-acetylcysteine potentiates inhibition of platelet aggregation by nitroglycerin. *J Clin Invest* 1985;76:703–708.
31. Ahlner J, Andersson RGG, Axelsson KL, et al. Effects of dipyridamole on the glyceryl-trinitrate-induced inhibition of coronary artery muscle tone and platelet aggregation in relation to cyclic nucleotide metabolism. *Acta Pharmacol Toxicol* 1985;57:88–95.
32. Chirkov YY, Naujalis Ji, Barber S, et al. Reversal of human platelet aggregation by low concentrations of nitroglycerin in vitro in normal subjects. *Am J Cardiol* 1992;70:802–806.
33. Chirkov YY, Belushkina NN, Tyshchuk IA, et al. Increase in reactivity of human platelet guanylate cyclase during aggregation portentiates the disaggregating capacity of sodium nitroprusside. *Clin Exp Pharmacol Physiol* 1991;18:517–524.
34. De Caterina R, Giannessi D, Crea F, et al. Inhibition of platelet function by injectable isosorbide dinitrate. *Am J Cardiol* 1984;53:1683–1687.
35. De Caterina R, Lombardi M, Bernini W, et al. Inhibition of platelet function during in vivo infusion of isosorbide

mononitrates: Relationship between plasma durg concentration and hemodynamic effects. *Am Heart J* 1990;119:855–862.

36. Diodati J, Théroux P, Latour JG, et al. Effects of nitroglycerin at therapeutic doses on platelet aggregation in unstable angina pectoris and acute myocardial infarction. *Am J Cardiol* 1990;66:683–688.
37. Karlberg K-E, Torfgård K, Ahlner J, et al. Dose-dependent effect of intravenous nitroglycerin on platelet aggregation, and correlation with plasma glyceryl dinitrate concentration in healthy men. *Am J Cardiol* 1992;69:802–805.
38. Lam JYT, Chesebro JH, Fuster V. Platelets, vasoconstriction, and nitroglycerin during arterial wall injury. *Circulation* 1988;78:712–716.
39. Brodie A, Chuck L, Klausner S, et al. Effects of sodium nitroprusside and nitroglycerin on tension prolongation of cat papillary muscle during recovery from hypoxia. *Circ Res* 1976;39:596–601.
40. Barath P, Kekes E, Strausz J. Acute global ischaemia and nitroglycerin-induced changes of the contraction-relaxation and diastolic properties of isolated rabbit heart. *Acta Physiol Pol* 1981;32:637–649.
41. Strauer BE, Westberg C, Tauchert M. Untersuchungen Über inotrope Nitroglycerinwirkungen am isolierten Ventrikelmyokard. *Pflügers Arch* 1971;324:124–133.
42. Himori N, Imai Y, Taira N. Positive inotropic action of glyceryl trinitrate as observed in the blood-perfused papillary muscle preparation of the dog heart. *Clin Exp Pharmacol Physiol* 1977;4:257–262.
43. Atanassova R, Spassov G, Balansky R, et al. Effects of isosorbide-5-mononitrate and isosorbide-2-mononitrate on the contractile and electrical activity and on the content of cyclic nucleotides in isolated heart muscles of the guineapig and dog. *J Pharm Pharmacol* 1992;44:663–666.
44. Laustiola K, Vuorinen P, Vapaatalo H, et al. Effects of cysteine and nitroglycerin on bovine heart guanylate cyclase and on tissue cyclic GMP and lacate of rat atria. *Eur J Pharmacol* 1983;91:301–304.
45. Ekstam Ljusegren M, Asplund A, Axelsson KL. Oxygen consumption in isolated hypoxic rat ventricular myocardium. Effect of different modulators of the cyclic GMP system. *Pharmacol Toxicol*, 1993, submitted.
46. Ahlner J. Andersson RGG, Torfgård K, et al. Organic nitrate esters: Clinical use and mechanisms of actions. *Pharmacol Rev* 1991;43:351–423.
47. Iversen HK. N-acetylcystein enhances nitroglycerin-induced headache and cranial arterial response. *Clin Pharmacol Ther* 1992;52:125–133.
48. Torfgård K, Ahlner J, Axelsson KL, et al. Tissue distribution of glyceryl trinitrate and the effect on cGMP levels in rat. *Pharmacol Toxicol* 1989;64:369–372.
49. Marks GS, McLaughlin BE, Nakatsu K, et al. Biotransformation of glyceryl trinitrate by rat brain homogenate. *Can J Physiol Pharmacol* 1992;70:935–937.
50. Kaverina NV. The mechanism of nitroglycerin action on the cardiac vessels. *Bull Exp Biol Med* 1960;49:75–78.
51. Kaverina NV, Vysotskaya NB, Rosonov YUB, et al. Connection between central action of glyceryl trinitrate and monoamine metabolism of the brain. *Bull Exp Biol Med* 1967;64:1310–1313.
52. Ma SX, Long JP. Central noradrenergic activity is responsible for nitroglycerin-induced cardiovascular effects in the nucleus tractus solitarii. *Brain Res* 1991;559;297–303.

53. Ma SX, Long JP, Bhatnagar RK. Further studies on central actions of nitroglycerin and lack of evidence for nitroglycerin interacting on [^{3}H]clonidine binding sites in cortex membranes. *J Pharmacol Exp Ther* 1992;261:1187–1194.
54. Radomski MW, Palmer RMJ, Moncada S. Endogenous nitric oxide inhibits human platelet adhesion to vascular endothelium. *Lancet* 1987;2:1057–1058.
55. Kubes P, Suzuki M, Granger DN. Nitric oxide: An endogenous modulator of leukocyte adhesion. *Proc Natl Acad Sci USA* 1991;88:4651–4655.
56. Boulanger C, Lüscher TF. Release of endothelin from the porcine aorta. Inhibition by endothelium-derived nitric oxide. *J Clin Invest* 1990;85:587–590.
57. Nakai T, Nakayama M, Kato R. Inhibition by nitric oxide and nitric oxide-producing vasodilators of DNA synthesis in vascular smooth muscle cells. *Eur J Pharmacol* 1990;1898:347–353.
58. Yates MT, Lambert LE, Whitten JP, et al. A protective role for nitric oxide in the oxidative modification of low density lipoproteins by mouse macrophages. *FEBS Lett* 1992;309:135–138.
59. Duarte IDG, Dos Santos IR, Lorenzetti BB, et al. Analgesia by direct antagonism of nociceptor sensitization involves the arginine-nitric oxide-cGMP pathway. *Eur J Pharmacol* 1992;217:225–227.
60. Ferreira SH, Lorenzetti BB, Faccioli LH. Blockade of hyperalgesia and neurogenic oedema by topical application of nitroglycerin. *Eur J Pharmacol* 1992;217:207–209.
61. Needleman P, Jakschik B, Johnson EM. Sulfhydryl requirement for relaxation of vascular smooth muscle. *J Pharmacol Exp Ther* 1973;187:324–331.
62. Schrör K, Grodzinska L, Darius H. Stimulation of coronary vascular prostacyclin and inhibition of human platelet thromboxane A_2 after low-dose nitroglycerin. *Thromb Res* 1981; 23:59–67.
63. Levin RI, Jaffe EA, Weksler BB, Tack-Goldman K. Nitroglycerin stimulates synthesis of prostaglandin by cultured human endothelial cells. *J Clin Invest* 1981;67:762–769.
64. Bennett BM, Moffat JA, Armstrong PW, Marks GS. Investigation of the role of prostaglandins in nitroglycerin-induced relaxation of isolated rabbit blood vessels. *Can J Physiol Pharmacol* 1983:61:554–560.
65. Rehr RB, Jackson JA, Winniford MD, et al. Mechanism of nitroglycerin-induced coronary dilatation. Lack of relation to intracoronary thromboxane concentrations. *Am J Cardiol* 1984;54:971–974.
66. DeCaterina R, Dorso CR, Tack-Goldman K, Weksler BB. Nitrates and endothelial prostacyclin production: Studies in vitro. *Circulation* 1985;71:176–182.
67. Bornfeldt KE, Axelsson KL. Studies on the effect of different inhibitors of arachidonic acid metabolism on glyceryltrinitrate-induced relaxation and cGMP elevation in bovine vascular tissue. *Pharmacol Toxicol* 1987;60:110–116.
68. Malta E. Studies on the biphasic relaxant curves of glyceryltrinitrate in rat aorta: Role of GTN metabolism. *Clin Exp Pharmacol Physiol* 1989;16:829–836.
69. Hamaguchi M, Ishibashi T, Imai S. Involvement of charybdotoxin-sensitive K^+ channel in the relaxation of bovine tracheal smooth muscle by glyceryl trinitrate and sodium nitroprusside. *J Pharmacol Exp Ther* 1992;262:263–270.
70. Williams DL, Katz GM, Roy-Contancin L, et al. Guanosine 5′-monophosphate modulates gating of high-conductance Ca^{2+}-

activated K^+ channels in vascular smooth muscle cell. *Proc Natl Acad Sci USA* 1988;85:9360–9364.

71. Fujino K, Nakaya S, Wakatsuki T, et al. Effects of nitroglycerin on ATP-induced Ca^{++}-mobilization, Ca^{++}-activated K channels and contraction of cultured smooth muscle cells of porcine coronary artery. *J Pharmacol Exp Ther* 1991; 256:371–377.
72. Imai S, Takeda K. Effect of vasodilators upon the isolated taenia coli of the guinea pig. *J Pharmacol Exp Ther* 1967;156:557–564.
73. Ito Y, Kitamura K, Kuriyama H. Actions of nitroglycerin on the membrane and mechanical properties of smooth muscles of the coronary artery in the pig. *Br J Pharmacol* 1980;70:197–204.
74. Karashima S. Actions of nitroglycerin on smooth muscles of guinea-pig and rat portal veins. *Br J Pharmacol* 1980; 71:489–497.
75. Tare M, Parkington HC, Coleman HA, et al. Hyperpolarisation and relaxation of arterial smooth muscle caused by nitric oxide derived from the endothelium. *Nature* 1990;346:69–71.
76. Stiefel A, Kreye VAW. On the hemodynamic differences between sodium nitroprusside, nitroglycerin, and isosorbide dinitrates. Comparison of their vasorelaxant effects in vitro and their inactivation in vivo. *Naunyn Schmiedebergs Arch Pharmacol* 1984;325;270–274.
77. Schneeweiss A. Comparative evaluation of isosorbide-5-mononitrate and nitroglycerin in chronic congestive heart failure. *Am J Cardiol* 1988;61:19E–21E.
78. Schneider W, Tessmer G, Strohm WD, et al. Die Weite von arteriellen und venösen Abdominalgefässen unter akuter und chronischer Gabe von Nitraten. Eine sonographische Untersuchung. *Z Kardiol* 1986;75:296–302.
79. Roth A, Weber L, Friedenberger L, et al. Hemodynamic effects of intravenous isosorbide dinitrate and nitroglycerin in acute myocardial infarction and elevated pulmonary artery wedge pressure. *Chest* 1987;91:190–196.
80. Bennet BM, Hayward LD, Murad F. Effects of the D and L stereoisomers of isosorbide dinitrate on relaxation and cyclic GMP accumulation in rat aorta and comparison to glyceryl trinitrate. *J Appl Cardiol* 1986;1:200–209.
81. Bennet BM, Schröder H, Hayward LD, et al. Effect of in vitro organic nitrate tolerance on relaxation, cyclic GMP accumulation, and guanylate cyclase activation by glyceryl trinitrate and the enantiomers of isosorbide dinitrate. *Circ Res* 1988;63:693–701.
82. Axelsson KL, Anderson C, Ahlner J, et al. Comparative in vitro study of a series of organic nitroesters: Unique biphasic concentration-effect curves for glyceryl trinitrate in isolated bovine arterial smooth muscle and lack of stereo-selectivity for some glyceryl trinitrate analogues. *J Cardiovasc Pharmacol* 1992;19:953–957.
83. Kamitani T. Inhibitory effect of Ca^{2+} on nitroglycerin-induced vasodilatation. *Eur J Pharmacol* 1984;100:145–154.
84. Ahlner J, Andersson RGG, Axelsson KL, et al. Development of tolerance towards glyceryl trinitrate in vitro. Low concentration induced tolerance and changes in cyclic guanosine-3′,5′-monophosphate metabolism. *Arzneim Forsch Durg Res* 1987;37:914–916.
85. Malta E. Biphasic relaxant curves to glyceryl trinitrate in rat aortic rings. Evidence for two mechanisms of action. *Naunyn Schmiedebergs Arch Pharmacol* 1989;339:236–243.
86. Axelsson KL, Ahlner J, Torfgård K. Effects of pertussis toxin and the cGMP 1 lowering agent LY83583 on the relaxation

induced by nitrates in isolated bovine mesenteric artery. A comparison between glyceryl trinitrate, ISDN and isosorbide-5 mononitrate. *Z Kardiol* 1989;78(Suppl 2):11–13.

87. Ignarro LJ, Lippton H, Edwards JC, et al. Mechanism of vascular smooth muscle relaxation by organic nitrates, nitrites, nitroprusside and nitric oxide: Evidence for the involvement of S-nitrosorthiols as active intermediates. *J Pharmacol Exp Ther* 1981;218:739–749.
88. Fung H-L, Chong S, Kowaluk E, et al. Mechanisms for the pharmacologic interaction of organic nitrates with thiols. Existence of an extracellular pathway for the reversal of nitrate vascular tolerance by N-acetylcystein. *J Pharmacol Exp Ther* 1988;245:524–530.
89. Chung S-J, Fung H-L. Identification of the subcellular site for nitroglycerin metabolism to nitric oxide in bovine coronary smooth muscle cells. *J Pharmacol Exp Ther* 1990;253:614–619.
90. Brien JF, McLaughlin BE, Kubos SM, et al. Mechanism of glyceryl trinitrate-induced vasodilation. I. Relationship between drug biotransformation, tissue cyclic GMP elevation and relaxation of rabbit aorta. *J Pharmacol Exp Ther* 1988;244:322–327.
91. Noonan PK, Benet LZ. Formation of mono- and dinitrate metabolities of nitroglycerin following incubation with human blood. *Int J Pharm* 1982;12:331–340.
92. Cossum PA, Roberts MS. Nitroglycerin Disposition in human blood. *Eur J Clin Pharmacol* 1985;29:169–175.
93. Tam GS, Marks G, Brien JF, Nakatsu K. Sex- and species-related differences in the biotransformation of isosorbide dinitrate by various tissues of the rabbit and rat. *Can J Physiol Pharmacol* 1987;65:1478–1483.
94. Chong S, Fung H-L. Thiol-mediated catalysis of nitroglycerin degradation by serum proteins. Increase in metabolism was not accompanied by S-nitrosothiol production. *Drug Metab Dispos* 1990;18:61–67.
95. Cossum PA, Robers MS, Yong AC, Kilpatrick D. Distribution and metabolism of nitroglycerin and its metabolites in vascular beds of sheep. *J Pharmacol Exp Ther* 1986;237:959–966.
96. Tam GS, MacMillan H, Bennett BM, Marks GS, Brien JF, Nakatsu K. Patterns of isosorbide dinitrate and glyceryl trinitrate metabolites formed by selected segments of rabbit gastrointestinal tract. *Can J Physiol Pharmacol* 1988;66:166–170.
97. Sellke FW, Tomanek RJ, Harrison DG. L-cysteine selectivity potentiates nitroglycerin induced dilation of small coronary microvessels. *J Pharmacol Exp Ther* 1991;258:365–369.
98. Chong S, Fung H-L. Kinetic mechanisms for the concentration dependency of in vitro degradation of nitroglycerin and glyceryl dinitrates in human blood. Metabolite inhibition or co-substrate depletion? *J Pharm Sci* 1989;78:295–302.
99. Hill KE, Hunt RW, Jones R, et al. Metabolism of nitroglycerin by smooth muscle cells. *Biochem Pharmacol* 1992;43:561–566.
100. Chung S-J, Chong S, Seth P, et al. Conversion of nitroglycerin to nitric oxide in microsomes of the bovine coronary artery smooth muscle is not primarily mediated by glutathione-S-transferases. *J Pharmacol Exp Ther* 1992;260:652–659.
101. Chung S-J, Fung H-L. A common enzyme may be responsible for the coversion of organic nitrates to nitric oxide in vascular microsomes. *Biochem Biophys Res Commun* 1992;185:932–937.

102. Seth P, Fung H-L. Biochemical characterization of a membrane bound enzyme responsible for generating nitric oxide from nitroglycerin in vascular smooth muscle cells. *Biochem Pharmacol* 1993;46:1481–1486.
103. Sakanashi M, Matsuzaki T, Aniya Y. Nitroglycerin relaxes coronary artery of the pig with no change in glutathione content or glutathione S-transferase activity. *Br J Pharmacol* 1991;103:1905–1908.
104. Chung S-J, Fung H-L. Relationship between nitroglycerin-induced vascular relaxation and nitric oxide production. Probes with inhibitors and tolerance development. *Biochem Pharmacol* 1993;45:157–163.
105. Lau DT-W, Benet LZ. Nitroglycerin metabolism in subcellular fractions of rabbit liver. Dose dependency of glyceryl dinitrate formation and possible involvement of multiple enzymes of glutathione S-transferases. *Drug Metab Dispos* 1990;181:489–497.
106. Marcus CJ, Habig WH, Jakobi WB. Glutathione transferase from human erythrocytes. *Arch Biochem Biophys* 1978;188:287–293.
107. Tsuchida S, Maki T, Sato K. Purification and characterization of glutathione transferase with an activity toward nitroglycerin from human aorta and heart. *J Biol Chem* 1990; 265:7150–7157.
108. Lau DT-W, Benet LZ. Effect of sulfobromophtalein and ethacrynic acid on glyceryl trinitrate relaxation. *Biochem Pharmacol* 1992;43:2247–2254.
109. Yeates RA, Schmid M, Leitold M. Antagonist of glyceryl trinitrate activity by inhibitors of glutathione-*S*-transferase. *Biochem Pharmacol* 1989;38:1749–1753.
110. Nigam R, Whiting T, Bennett BM. Effect of glutathione S-transferase on glyceryl trinitrate actvity in isolated rat aorta. *Can J Physiol Pharmacol* 1993;71:179–184.
111. Servent D, Delaforge M, Ducrocq C, et al. Nitric formation during microsomal hepatic denitration of glyceryl trinitrate: Involvement of cytochrome P-450. *Biochem Biophys Res Commun* 1989;163:1210-1216.
112. McDonald BJ, Bennett BM. Cytochrome P-450 mediated biotransformation of organic nitrates. *Can J Physiol Pharmacol* 1990;68:1552–1557.
113. Abraham NG, Pinto A, Mullane KM, et al. Presence of cytochrome P-450-dependent monooxygenase in intimal cells of the hog aorta. *Hypertension* 1985;7:899–904.
114. Liu Z, Brien F, Marks GS, et al. Lack of evidence for the involvement of cytochrome P-450 or other hemoproteins in metabolic activations of glyceryl trinitrate in rabbit aorta. *J Pharmacol Exp Ther* 1993;264:1432–1439.
115. Bennett BM, McDonald BJ, Nigam R, et al. Inhibition of nitrovasodilator- and acetylcholine-induced relaxation and cyclic GMP accumulation by the cytochrome P-450 substrate, 7-ethoxyresorufin. *Can J Physiol Pharmacol* 1993;70:1297–1303.
116. Schröder H, Schrör K. Inhibitors of cytochrome P-450 reduce cyclic GMP stimulation of glyceryl trinitrate in $LLCPK_1$ kidney epithelial cells. *Naunyn-Schmiedebergs Arch Pharmacol* 1990;342:616–618.
117. Schröder H. Cytochrome P-450 mediates bioactivation of organic nitrates. *J Pharmacol Exp Ther* 1992;262:298–302.
118. Bennett BM, McDonald BJ, St. James MJ. Hepatic cytochrome P-450-mediated activation of rat aortic guanylyl cyclase by glyceryl trinitrate. *J Pharmacol Exp Ther* 1992;261:716–723.

119. Williams FM. Clinical significance of esterases in man. *Clin Pharmacokin* 1985;10:392–403.
120. Noak E. Investigation on structure-activity relationship in organic nitrates. *Meth Exp Clin Pharmacol* 1984;6:583–586.
121. Chhajlani V, Axelsson KL, Ahlner J, et al. Purification of soluble guanylate cyclase enzyme from human platelets. *Biochem Int* 1989;19:1039–1044.
122. Kamisaki Y, Saheki S, Nakane M, et al. Soluble guanylate cyclase from rat lung exists as a heterodimer. *J Biol Chem* 1986;261:7236–7241.
123. Koesling D, Herz J, Gausepohl H, et al. The primary structure of the 70 kDa subunit of bovine soluble guanylate cyclase. *FEBS Lett* 1988;239:29–34.
124. Nakane M, Arai K, Saheki S, et al. Molecular cloning and expression of cDNAs coding for soluble guanylate cyclase from rat lung. *J Biol Chem* 1990;265:16841–16845.
125. Buechler WA, Nakane M, Murad F. Expression of soluble guanylate cyclase activity requires both enzyme subunits. *Biochem Biophys Res Commun* 1991;174:351–357.
126. Gerzer R, Böhme E, Hofmann F, et al. Soluble guanylate cyclase purified from bovine lung contains copper and heme. *FEBS Lett* 1981;132:71–74.
127. Gerzer R, Radany EW, Garbers DL. The separation of the heme and apoheme forms of soluble guanylate cyclase. *Biochem Biophys Res Commun* 1982;108:678–686.
128. Kon H, Kataoka N. Electron paramagnetic resonance of nitric oxide-protoheme complexes with some nitrogenous base. Model systems of nitric oxide hemoproteins. *Biochemistry* 1969;8:4757–4762.
129. Morse RH, Chan SI. Electron paramagnetic resonance studies of nitrosyl ferrous heme complexes. Determination of an equilibrium between two conformations. *J Biol Chem* 1980;255:7876–7882.
130. Wolin MS, Wook KS, Ignarro LJ. Guanylate cyclase from bovine lung. A kinetic analysis of the regulation of the unpurified soluble enzyme by protoporphyrin IX, heme, and nitrosyl-heme. *J Biol Chem* 1982;257:13312–13320.
131. Ignarro LJ, Wood KS, Wolin KS. Regulation of purified soluble guanylate cyclase by porphyrins and metalloporphyrins: A unifying concept. *Adv Cycl Nucl Prot Phosphoryl Res* 1984;17:267–274.
132. Craven PA, DeRubertis FR. Effects of thiol inhibitors on hepatic guanylate cyclase activity. Evidence for the involvement of vicinal dithiols in the expression of basal and agonist-stimulated activity. *Biochim Biophys Acta* 1978a; 524:231–244.
133. Ignarro LJ, Kadowitz PJ, Baricos WH. Evidence that regulation of hepatic guanylate cyclase activity involves interactions between catalytic site-SH groups and both substrate and activator. *Arch Biochem Biophys* 1981;208;75–86.
134. Heppel LA, Hilmoe RJ. Metabolism of inorganic nitrite and nitrate esters. II. The enzymatic reduction of nitroglycerin and erythritoltetranitrate by glutathione. *J Biol Chem* 1950;183:129–138.
135. Ignarro LJ, Edwards JC, Gruetter DY, et al. Possible involvement of S-nitrosothiols in the activation guanylate cyclase by nitroso compounds. *FEBS Lett* 1980;110:275–278.
136. Ignarro LJ, Gruetter CA. Requriements of thiols for activation of coronary arterial guanylate cyclase by glyceryl trinitrate and sodium nitrite: Possible involvement of S-nitrosothiols. *Biochem Biophs Acta* 1980;631:221–231.

137. Craven PA, DeRubertis FR. Restoration of responsiveness of purified guanylate cyclase to nitrosoguanidine, nitric oxide, and related activators by heme and hemeproteins. *J Biol Chem* 1978;253:8433–8443.
138. Craven PA, DeRubertis FR. Requirement for heme in the activation of purified guanylate cyclase by nitric oxide. *Biochem Biophys Acta* 1983;745:310–321.
139. Beavo JA, Reifsnyder DH. Primary sequence of cyclic nucleotide phosphodiesterase isoenzymes and the design of selective inhibitors. *Trends Pharmacol Sci* 1990;11:150–155.
140. Weishaar RE, Burrows SD, Kobylarz DC, et al. Multiple forms of cyclic nucleotide phosphodiesterase in cardiac and smooth muscle in platelets. Isolation, characterization, and effects of various reference phosphodiesterase inhibitors and cardiotonic agents. *Biochem Pharmacol* 1986;35:787–800.
141. Gillespie PG, Beavo JA. Inhibition and stimulation of photoreceptor phosphodiesterases by dipyridamole and M&B 22,948. *Mol Pharmacol* 1989;36:773–781.
142. Sutherland EW, Robison GA, Butcher RW. Some aspects of the biological role of adenosine-3′,5′-monophosphate (cyclic AMP). *Circulation* 1968;37:279–306.
143. Kobayashi A, Suzuki Y, Kamikawa T, et al. The effects of nitroglycerin on cyclic nucleotides in the coronary artery in vivo. *Life Sci* 1980;27:1679–1685.
144. Kukovetz WR, Holzmann S, Wurm A, et al. Evidence for cyclic GMP-mediated relaxant effects of nitro-compounds in coronary smooth muscle. *Naunyn Schmiedebergs Arch Pharmacol* 1979;310:129–138.
145. Holzmann S. Endothelium-induced relaxation by acetylcholine associated with larger rises in cyclic GMP in coronary arterial strips. *J Cycl Nucl Prot Phosphoryl Res* 1982;8:409–419.
146. Ahlner J, Andersson RGG, Axelsson KL, et al. Development of tolerance to glyceryl trinitrate in an isolated human peripheral vein and its relation to cyclic GMP metabolism. *Acta Pharmacol Toxicol* 1986;59:123–128.
147. Schultz K-D, Böhme E, Kreye VAW, et al. Relaxation of hormonally stimulated smooth muscular tissues by the 8-bromoderivative of cyclic GMP. *Naunyn Schmiedebergs Arch Pharmacol* 1979;306:1–9.
148. Napoli SA, Gruetter CS, Ignarro LJ, et al. Relaxation of bovine coronary arterial smooth muscle by cyclic GMP, cyclic AMP and analogs. *J Pharmacol Exp Ther* 1980;212:469–473.
149. Axelsson KL. Studies on the Action of Organic Nitroesters in Vascular Smooth Muscle with Special Reference to Cyclic GMP Metabolism. Linköping University Medical Dissertations no. 179, Linköping 1984, pp 1–58.
150. Saeed M, Schmidli J, Metz M, et al. The effect of 8-bromo-cyclic GMP on rabbit vascular smooth muscle. *Curr Ther Res* 1987;41:199–209.
151. Axelsson KL, Karlsson JOG. Nitroglycerin tolerance in vitro: Effect on cGMP turnover in vascular smooth muscle. *Acta Pharmacol Toxicol* 1984;55:203–210.
152. Romanin C, Kukovetz WR. Guanylate cyclase activation by organic nitrates is not mediated via nitrite. *J Mol Cell Cardiol* 1988;20:389–396.
153. Milligan G. Techniques used in the identification and analysis of function of pertussis toxin-sensitive guanine nucleotide binding proteins. *Biochem J* 1988;255:1–13.
154. Reisine T. Pertussis toxin in the analysis of receptor mechanisms. *Biochem Pharmacol* 1990;39:1499–1504.

155. Ahlner J, Axelsson KL, Ekstam Ljusegren M, et al. Relaxation of bovine mesenteric artery induced by glyceryl trinitrate is attenuated by pertussis toxin. *Pharmacol Toxicol* 1988;62:155–158.
156. Lobban M, Van Heyningen S. Thiol reagents are substrates for the ADP-ribosyltransferase activity of pertussis toxin. *FEBS Lett* 1988;233;229–232.
157. Yuen PST, Potter LR, Garbers DL. A new form of guanylyl cyclase is preferentially expressed in rat kidney. *Biochemistry* 1990;29:10872–10878.
158. Chhajlani V, Frändberg PA, Ahlner J, et al. Heterogeneity in human soluble guanylate cyclase due to alternative splicing. *FEBS Lett* 1991;290:157–158.
159. Lad PJ, White AA. Activation of particulate guanylate cyclase by nitroprusside and MNNG after filipin treatment. *J Cycl Nucl Res* 1979;5:315–325.
160. Waldman SA, Lewicki JA, Brandwein HJ, et al. Partial purification and characterization of particulate guanylate cyclase from rat liver after solubilization with trypsin. *J Cycl Nucl Res* 1982;8:359–370.
161. Waldman SA, Murad F. Cyclic GMP synthesis and function. *Pharmacol Rev* 1987;39:163–196.
162. Epstein SA, Giannella RA, Brandwein HJ. Activation of guanylate cyclase by E. coli heat-stable enterotoxin (STa). *FEBS Lett* 1986;203:44–48.
163. Cornwell TL, Pryzwansky KB, Wyatt TA, Lincoln TM. Regulation of sarcoplasmic reticulum protein phosphorylation by localized cyclic GMP-dependent protein kinase in vascular smooth muscle cells. *J Pharmacol Exp Ther* 1991;40:923–931.
164. Stamler J, Cunningham M, Loscalzo J. Reduced thiols and the effect of intravenous nitroglycerin on platelet aggregation. *Am J Cardiol* 1988;62:377–380.
165. Takai Y, Kaibuchi K, Matsubara T, et al. Inhibitory action of guanosine 3′,5′-monophosphate on thrombin-induced phosphatidylinositol turnover and protein phosphorylation in human platelets. *Biochem Biophys Res Commun* 1981; 101;61–67.
166. Yamanishi J, Tajai Y, Kaibuchi K, et al. Synergistic functions of phorbol ester and calcium in serotonin release from human platelets. *Biochem Biophys Res Commun* 1983;112:778–786.
167. Radomski MW, Palmer RMJ, Moncada S. The role of nitric oxide and cGMP in platelet adhesion to vascular endothelium. *Biochem Biophys Res Commun* 1987;148:1482–1489.
168. Sneddon JM, Vane JR. Endothelium-derived relaxing factor reduces platelet adhesion to bovine endothelial cells. *Proc Natl Acad Sci USA* 1988;85:2800–2804.
169. Groves PH, Penny WJ, Cheadle HA, et al. Exogenous nitric oxide inhibits in vivo platelet adhesion following balloon angioplasty. *Cardiovasc Res* 1992;26:615–619.
170. Venturini CM, Weston LK, Kaplan JE. Platelet cGMP, but not cAMP, inhibits thrombin-induced platelet adhesion to pulmonary vascular endothelium. *Am J Physiol* 1992;263:H606–H612.
171. Lieberman EH, O'Neill S. Mendelsohn ME. S-Nitrosocysteine inhibition of human platelet secretion is correlated with increases in platelet cGMP levels. *Circ Res* 1991;68:1722–1728.
172. Drummer C, Lüdke S, Spannagl M, et al. The nitric oxide donor SIN-1 is a potent inhibitor of plasminogen activator inhibitor release from stimulated platelets. *Thronmb Res* 1991;63:553–556.

173. Eigenthaler NM, Nolte C, Halbrügge M, et al. Concentration and regulation of cyclic nucleotides, cyclic-nucleotide-dependent protein kinases and one of their major substrates in human platelets. *Eur J Biochem* 1992;205:471–481.
174. Laustiola K, Vuorinen P, Karp M, et al. 8-Bromo cyclic GMP inhibits NADH and lactate accumulation in hypoxic rat atria. *Naunyn Schmiedebergs Arch Pharmacol* 1983;323:361–363.
175. Laustiola K, Vuorinen P, Vapaatalo H, et al. Sodium nitroprusside inhibits lactate formation in rat atria: Is cyclic GMP involved? *Acta Pharmacol Toxicol* 1983;52:195–200.
176. Vuorinen P, Laustiola K, Metsä-Ketelä. The effects of cyclic AMP and cyclic GMP on redox state and energy state in hypoxic rat atria. *Life Sci* 1984;35:155–161.
177. Laustiola K, Vuorinen P, Metsä-Ketelä T. 8-Bromo-cGMP improves energy state in hypoxic rat atria. *Acta Pharmacol Toxicol* 1984;55:21–24.
178. Laustiola K. Cyclic GMP affects redox state and improves energy charge of ischemic Langendorf-perfused rat heart. *Acta Pharmacol Toxicol* 1985;56:139–143.
179. Ekstam Ljusegren M, Axelsson KL. Lactate accumulation in isolated hypoxic rat ventricular myocardium. Effect of different modulators of the cyclic GMP system. *Pharmacol Toxicol* 1993;72:56–60.
180. Shah AM, Lewis MJ, Henderson AH. Effects of 8-bromo-cyclic GMP on contraction and on inotropic responses of ferret cardiac muscle. *J Mol Cell Cardiol* 1991;23:55–64.
181. Shah AM, Evans HG, Lewis MJ. Phenylephrine-induced cardiac phosphatidylinositol (PI) hydrolysis is inhibited by cyclic GMP. *Circulation* 1990;82(Suppl III):III-270.
182. Bulmenthal DK, Stull JT, Gill GN. Phosphorylation of cardiac troponin by guanosine 3′:5′-monophosphate-dependent kinase. *J Biol Chem* 1978;253:334–336.
183. Lincoln TM, Corbin JD. Purified cyclic GMP-dependent protein kinase catalyzes the phosphorylation of cardiac troponin inhibitory subunit (TN-I). *J Biol Chem* 1978; 253:337–339.
184. Clement O, Puceat M, Vassort G. Protein kinases modulate Ca sensitivity of cardiac myofilaments in rat skinned cells. *J Physiol (Lond)* 1991;438:96P.
185. Van Renterghem C, Vigne P, Barhemin J, et al. Molecular mechanism of action of the vasoconstrictor peptide endothelin. *Biochem Biophys Res Commun* 1988;157:977–985.
186. Hendersen AH, Lewis MJ, Shah AM, et al. Endothelium, endocardium, and cardiac contraction. *Cardiovasc Res* 1992;26:305–308.
187. Hofmann F, Dostmann W, Keilbach A, et al. Structure and physiological role of cGMP-dependent protein kinase. *Biochim Biophys Acta* 1992;1135:51–60.
188. Axelsson KL, Karlsson J-O G, Pettersson G. Effect of glyceryl trinitrate and 8-Br-cGMP on tension and phosphorylase activity in vascular smooth muscle. *Acta Pharmacol Toxicol* 1985;57:227–233.
189. Johnson RM, Lincoln TM. Effects of nitroprusside, glyceryl trinitrate, and 8-bromo cyclic GMP on phosphorylase a formation and myosin light chain phosphorylation in rat aorta. *Mol Pharmacol* 1985;27:333–342.
190. Lincoln TM, Cornwell TL, Rashatwar SS, et al. Mechanism of cyclic-GMP-dependent relaxation in vascular smooth muscle. *Biochem Soc Transact* 1988;16:497–499.
191. Kawahara Y, Yamanishi J, Fukuzaki H. Inhibitory action of guanosine 3′,5′-monophosphate on thrombin-induced cal-

cium mobilization in human platelets. *Thromb Res* 1984;33:203–209.

192. Lockette W, McCurdy R, Aronow, et al. Effect of 8-bromo-cyclic guanosine monophosphate on intracellular pH and calcium in vascular smooth muscle. *Hypertension* 1989;13:865–869.
193. Morgan RO, Newby AC. Nitroprusside differentially inhibits ADP-stimulated calcium influx and mobilization in human platelets. *Biochem J* 1989;258:447–454.
194. Negrescu EV, Sazonova LN, Baldenkov GN, et al. Relationship between the inhibition of receptor-induced increase in cytosolic free calcium concentration and the vasodilator effects of nitrates in patients with congestive heart failure. *Int J Cardiol* 1990;26:175–184.
195. Grün G, Fleckenstein A. Die elektromechanische Entkopplung der glatten Gefässmuskulatur als Grundprincip der Koronardilatation durch 4-(2′-nitrophenyl)-2,6-dimetyl-1, 4-dihydropyridin-3,5-dicarbonsäure-dimetyl ester (BAYa 10410, Nifedipine), *Arzneim Forsch* 1972;22:334–344.
196. Weder U, Grün G. Ca-antagonistische elektromechanische Entkopplung der glatten Gefässmuskulatur als Wirkungsprincip vasodilatorischer Nitroverbindungen. *Naunyn Schmiedebergs Arch Pharmacol* 1973;277:R88.
197. Zostér TT, Jacyk P, Wolchinsky C. The effect of vasodilators on calcium kinetics in blood vessels. *Arch Int Pharmacodyn* 1974;212:328–336.
198. Zostér TT, Henein NF, Wolchinsky C. The effect of sodium nitroprusside on the uptake and efflux of ^{45}Ca from rabbit and rat vessels. *Eur J Pharmacol* 1977;45:7–12.
199. Hester RK, Weiss GB, Fry WJ. Differing actions of nitroprusside and D-600 on tension and ^{45}Ca fluxes in canine renal arteries. *J Pharmacol Exp Ther* 1979;208:155–160.
200. Kreye VAW, Baron GD, Lüth BJ, et al. Mode of action of sodium nitroprusside on vascular smooth muscle. *Naunyn Schmiedebergs Arch Pharmacol* 1975;288:381–402.
201. Thorens S, Haeusler G. Effects of some vasodilators on calcium translocation in intact and fractionated vascular smooth muscle. *Eur J Pharmacol* 1979;54:71–91.
202. Ahlner J, Axelsson KL, Karzewski P, et al. Studies on the effect of glyceryl trinitrate and cyclic GMP on calcium turnover in bovine mesenteric artery. *Pharmacol Toxicol* 1990;66:277–282.
203. Nakazawa M, Imai S. Effect of nitroglycerin on Ca efflux in the coronary artery. *Jpn J Pharmacol* 1981;31:621–625.
204. Ozaki H, Shibata S, Kitano H, et al. A comparative study of the effect of nitroprusside and verapamil on human umbilical vessels. *Blood Vessels* 1981;18:321–329.
205. Lincoln TM. Effects of nitroprusside and 8-Br-cyclic GMP on the contractile activity in the rat aorta. *J Pharmacol Exp Ther* 1983;224:100–107.
206. Twort CHA, Van Breemen C. Cyclic guanosine monophosphate-enhanced sequestration of Ca^{2+} by sarcoplasmic reticulum in vascular smooth muscle. *Circ Res* 1988;62:961–964.
207. Nakashima S, Tohmatsu T, Hattori H, et al. Inhibitory action of cyclic GMP on secretion, polyphosphoinositide hydrolysis and calcium mobilization in thrombin-stimulated human platelets. *Biochem Biophys Res Commun* 1986;135:1099–1104.
208. Fujii K, Ishimatsu T, Kuriyama H. Mechanism of vasodilation induced by α-human atrial natriuretic polypeptide in

rabbit and guinea-pig renal arteries. *J Physiol* 1986;377:315–332.

209. Rapoport RM. Cyclic guanosine monophosphate inhibition of contraction may be mediated by through inhibition of phosphatidylinositol hydrolysis in rat aorta. *Circ Res* 1986;58:407–410.
210. Ahlner J, Axelsson KL, Karlsson J-O G, et al. Glyceryl trinitrate inhibits phosphatidyl inositol hydrolysis and protein kinase C activity in bovine mesenteric artery. *Life Sci* 1988;43:1241–1248.
211. Nishizuka Y. The role of protein kinase C in cell surface signal transduction and tumor promotion. *Nature* 1984;308:693–698.
212. Movesian MA, Nishikawa M, Adelstein RS. Phosphorylation of phospholamban by calcium-activated, phospholipid-dependent protein kinase. *J Biol Chem* 1984;259:8029–8032.
213. Chiu A, Bozarth JM, Forsythe MS, et al. Ca^{++} utilization in the constriction of rat aorta to stimulation of protein kinase C by phorbol dibutyrate. *J Pharmacol Exp Ther* 1987;242:934–939.
214. Hagiwara M, Inagaki M, Watanabe M, et al. Selective modulation of calcium-dependent myosin phosphorylation by novel protein kinase inhibitors, isoquinolinesulfonamide derivatives. *Mol Pharmacol* 1987;32:7–12.
215. Hirata M, Kohse KP, Chang C-H, et al. Mechanism of cyclic GMP inhibition of inositol phosphate formation in rat aorta segments and cultured bovine aortic smooth muscle cells. *J Biol Chem* 1990;265:1268–1273.
216. Lincoln TM. Pertussis toxin-sensitive and insensitive guanine nucleotide binding proteins (G-proteins) are not phosphorylated by cyclic GMP-dependent protein kinase. *Second Messeng Phosphoprot* 1991;13:99–109.
217. Popescu LM, Panoiu C, Hinescu M, et al. The mechanism of cGMP-induced relaxation in vascular smooth muscle. *Eur J Pharmacol* 1985;107:393–394.
218. Furukawa K-I, Nakamura H. Cyclic GMP regulation of the plasma membrane (Ca^{2+}-Mg^{2+})ATPase in vascular smooth muscle. *J Biochem* 1987;101:287–290.
219. Yoshida Y, Sun H-T, Cai J-Q, et al. Cyclic GMP-dependent protein kinase stimulates the plasma membrane Ca^{2+} pump ATPase of vascular smooth muscle via phosphorylation of a 240-kDa protein. *J Biol Chem* 1991;266:19819–19825.
220. Yanagisawa T, Kawada M, Taira N. Nitroglycerin relaxes canine coronary arterial smooth muscle without reducing intracellular Ca^{2+} concentrations measured with fura-2. *Br J Pharmacol* 1989;98:469–482.
221. Rapoport RM, Draznin MB, Murad F. Endothelium-dependent relaxation in rat aorta may be mediated through cyclic GMP-dependent protein phosphorylation. *Nature* 1983;306:174–176.
222. Rapoport RM, Draznin MB, Murad F. Endothelium-dependent vasodilator- and nitrovasodilator-induced relaxation may be mediated through cyclic GMP formation and cyclic GMP-dependent protein phosphorylation. *Trans Assoc Am Phys* 1983;96:19–30.
223. Rapoport RM, Draznin MB, Murad F. Sodium nitroprusside-induced protein phosphorylation in intact rat aorta is mimicked by 8-bromo cyclic GMP. *Proc Natl Acad Sci USA* 1982;79:6470–6474.
224. Draznin MB, Rapoport RM, Martinez GA, et al. Myosin light chain dephosphorylation in intact rat thoracic aorta treated

with sodium nitroprusside: Role of cyclic GMP. *Clin Res* 1983;31:466A.

225. McDaniel NL, Chen X-L, Singer HA, et al. Nitrovasodilators relax arterial smooth muscle by decreasing $[Ca^{21}]$i and uncoupling stress from myosin phosphorylation. *Am J Physiol* 1992;263:C461–C467.

226. Garg UC, Hassid A. Nitric oxide decreases cytosolic free calcium in Balb/c 3T3 fibroblasts by a cyclic GMP-independent mechanism. *J Biol Chem* 1991;266:9–12.

227. Dimmeler S, Lottspeich F, Brüne B. Nitric oxide causes ADP-ribosylation and inhibition of glyceraldehyde-3-phosphate dehydrogenase. *J Biol Chem* 1992;267:16771–16774.

Mechanisms of Nitrate Tolerance

Ho-Leung Fung and John Anthony Bauer

Summary. **There is now little dispute that clinical tolerance of organic nitrates occurs, particularly when these drugs are used by themselves to treat patients with stable angina pectoris and congestive heart failure. Classical hypotheses of nitrate tolerance suggest the phenomenon to result from vascular depletion of critical sulfhydryl groups, which are necessary to bring about vasorelaxation from nitrates. While this mechanism of nitrate tolerance probably operates when isolated blood vessels are exposed to high concentrations of nitrate in vitro, there is little evidence to suggest that it contributes to clinical nitrate tolerance. Instead, emerging data suggest that nitrates can cause significant shifts in fluid distribution and secretion of neurohormonal factors that can modulate their vasorelaxant effects. Use of angiotensin converting enzyme inhibitors and diuretics in conjunction with nitrates may alleviate the development of tolerance, but the experience has not been universally favorable. Other receptor-effector systems that affect cardiovascular function, such as the adrenergic system, may also be affected by nitrate tolerance. The mechanisms of nitrate tolerance are therefore likely to be multifactorial, involving vascular biochemical changes, physiologic compensation, and possibly receptor regulation.**

Key Words. **organic nitrates, nitroglycerin, tolerance, angina pectoris, congestive heart failure**

When given in acute doses, nitrates are safe, economic, and effective vasodilators that produce few debilitating side effects. However, their utility in the long-term management of cardiovascular diseases has been limited by the potential development of therapeutic tolerance that accompanies prolonged nitrate administration. The occurrence of clinical nitrate tolerance has been controversial until recent years, and convincing evidence of its existence was not well documented until the availability and widespread acceptance of the sustained-release nitroglycerin (NTG) transdermal patches in therapy. These dosage forms were designed to deliver NTG at a constant rate over 24 hours and therefore were anticipated to provide sustained prophylaxis against angina pectoris. Ironically, this constant delivery mode has made clinical nitrate tolerance more readily observable and documentable. Today this phenomenon has gained widespread acceptance.

Several reviews have appeared recently, describing the phenomenon of nitrate tolerance, its mechanisms, and methods that can be employed to combat it [1–3]. These reviews provide extensive literature citations, and readers are encouraged to refer to them. Here we shall not attempt to tread over similar grounds, but instead shall address only selected issues of nitrate tolerance, particularly including those that have been examined by our laboratory.

The Tolerance Phenomenon: How Solid is the Evidence?

In vitro vascular tolerance

There is little controversy about the occurrence of vascular tolerance in vitro when millimolar concentration of NTG were in contact with isolated vessel strips for 30 minutes or longer. The shift in the dose-response curve after tolerance development is substantial, with the sensitivity reduced at least two orders of magnitude at around the EC_{50}. Blood vessels that are tolerant toward NTG are crosstolerant to other organic nitrate esters. Incorporation of a drug-free period restores vascular reactivity toward NTG [4].

Clinical nitrate tolerance

The occurrence of clinical nitrate tolerance is now well recognized [1–3]. This phenomenon is not restricted to a particular cardiovascular disease, one of the nitrates, or a special dosage form or route of administration. Thus in patients with congestive heart failure, tolerance to the beneficial hemodynamic effects of nitrates have been shown toward intravenous NTG [5,6], transdermal NTG [7,8], and oral isosorbide dinitrate (ISDN) [9]. In patients with stable angia pectoris, tolerance to the antiischemic effects (which are most often measured in terms of improvements of the duration of exercise) of intravenous NTG [10], transdermal NTG [11–13], oral ISDN [14,15], transdermal ISDN [16], and oral isosorbide-5-mononitrate [17] have been well demonstrated.

While nitrates are often used in the treatment of unstable angina [18,19], the data on nitrate tolerance in this disease state are less available, and far from definitive. Whether continuous nitrate administration will induce tolerance in its antiplatelet effects [20] is still unanswered. Thus in vivo nitrate tolerance is well substantiated in patients with stable angina pectoris and congestive heart failure but is at present not sufficiently documented in patients with unstable angina.

Is in vitro nitrate tolerance arterial, venous, or both?

While nitrates are commonly thought to be primarily venodilators, they can and do exert substantial arterial effects under clinical doses. A natural question that can be asked is whether nitrate tolerance in vivo is differentially expressed in the venous versus arterial circulations. Data have been gathered both in experimental and clinical settings in an attempt to answer this question. Table 1 summarizes the pertinent information from these studies.

In an earlier clinical study, Zelis and Mason [21] gave sublingual NTG (0.9 mg) to six normal subjects before and after a treatment regimen of 6–8 weeks of isosorbide dinitrate (ISDN) three times daily, and measured blood pressure, calf blood flow, calf vascular resistance, and calf venous volume in these subjects. These authors showed that there was crosstolerance between ISDN and NTG, and that while the arteriolar dilation produced by NTG was unaltered by chronic ISDN dosing, the venodilation was significantly attenuated. The presence of a larger venous tolerance component was supported in two other clinical studies involving different patient populations. Makhoul et al. [22] administered an intravenous dose of NTG (525 ± 548 μg/min for 24 hours) to nine patients with congestive heart failure and showed that tolerance had a different time course and magnitude in the venous, arterial, and pulmonary circulations. After 24 hours of infusion, the initial hemodynamic effects of NTG on right atrial pressure and pulmonary vascular resistance were lost, while about half of the effects on systemic vascular resistance, cardiac index, and pulmonary wedge pressure were maintained. In a more recent study, Ghio et al. [23] examined 20 patients with coronary artery disease and measured the forearm vascular resistance and venous volume after 24 hours of intravenous NTG infusion at a dose of 169 ± 115 μg/min. These workers found that the initial increase in venous volume was reduced 48% after 24 hours of infusion, while the acute effects on vascular resis-

Table 1. *Is nitrate tolerance arterial, venous, or both?*

Study	Nitrate dose	Population	Study design	Arterial parameter(s)	Venous parameter(s)	Tolerance: Arterial vs. venous
A. Clinical studies						
Zelis and Mason [21]	NTG sl, 0.9 mg, before/after ISDN 40 mg tid, 6–8 weeks	6 normal subjects	sl challenge of NTG before and after ISDN for 6–8 weeks	Systemic BP, calf blood flow, calf VR	Calf venous volume	Venous > arterial tolerance
Manyari et al. [24]	NTG sl, 0.6 mg, before/after ISDN 4 weeks	10 patients with stable angina	sl challenge of NTG before and after ISDN for 4 weeks	Systolic BP	Regional blood volume	Both arterial and venous tolerance
Ghio et al. [23]	NTG IV, 169 ± 115 μg/min, 24 hr	20 patients with CAD	Continuous NTG IV, strain gauge plethysmography	Forearm VR	Forearm venous volume	Venous > arterial tolerance
Leier et al. [25]	ISDN 40 mg, q6h, 12 weeks	30 patients with CHF	R, DB, PC	BP, SVR	PCWP	Arterial > venous tolerance
Makhoul et al. [22]	NTG IV, 525 ± 548 μg/min, 24 hr	9 patients with CHF	NTG IV at max. tolerated dose for 24 hr	SVR, CI	RAP, PCWP	Venous ≥ arterial tolerance
Packer et al. [6]	NTG IV, 6.4 μg/kg/min, 48 hr	8 patients with severe CHF	Continuous NTG IV	MAP, SVI, SVR	PCWP, RAP	Both arterial and venous tolerance
Dupuis et al. [37]	NTG IV, 1.5 μg/kg/min, 24 hr	13 patients with CHF	Continuous NTG IV	MAP, SVI, SVR	PCWP, RAP	Both arterial and venous tolerance
B. Experimental studies						
Goldberg et al. [27]	NTG IV, 5 μg/kg/min, 2 hr	6 dogs with autonomic blockade, splenectomized	NTG IV for 2 hr, under anesthesia	MAP, arterial compliance	Venous compliance	Both arterial and venous tolerance in 2 hr
Stewart et al. [26]	NTG IV, 1.5 μg/kg/min, 5 days	6 dogs with autonomic blockade, anesthetized	NTG dose response before/ after 4 days NTG IV	MAP, SVR	Total effective vascular compliance	Venous > arterial tolerance

td, transdermal; IV, intravenous; sl, sublingual; NTG, nitroglycerin; ISDN, isosorbide dinitrate; NAC, N-acetyl cysteine; MET, methionine; CAD, coronary artery disease; CHF, congestive heart failure; R, randomized; DB, double blind; C-O, cross-over; PC, placebo controlled; PCWP, pulmonary capillary wedge pressure; MRAP, mean right atrial pressure; BP, blood pressure; HR, heart rate; SVI, stroke volume index; SVR, systemic vascular resistance; PAP, pulmonary artery pressure.

tance were not diminished, indicating that nitrate tolerance in humans is higher in the venous than in the arterial circulation.

Two other studies, however, did not arrive at the same conclusion. Manyari et al. [24] examined 10 patients with stable angina pectoris using sublingual NTG (0.6 mg) before and after 4 weeks of oral ISDN therapy. Using systolic blood pressure and reginal blood volume as arterial and venous indices, respectively, these authors concluded that both arterial and venous tolerance occurred. Leier et al. [25] studied 30 patients with congestive heart failure after 12 weeks of ISDN therapy at 40 mg every 6 hours, and concluded that arterial tolerance in blood pressure and systemic vascular resistance was more predominant than venous tolerance, as reflected by changes in the pulmonary capillary wedge pressure.

Stewart et al. [26] administered intravenous NTG to six dogs with autonomic blockade at a dose of 1.5 μg/kg/min for 5 days, and measured mean arterial pressure and total effective vascular compliance as arterial and venous parameters, respectively. These workers conclude that this exposure to NTG resulted in tolerance to its venodilating effects, while arteriolar action was maintained. In addition, lowering of mean arterial pressure was observed only after autonomic blockade. Goldberg et al. [27] used a higher NTG infusion rate (5 μg/kg/min for 2 hours) in splenectomized glanglion-blocked dogs and showed that both arterial and venous tolerance occurred within 2 hours.

The picture that emerges, therefore, is far from crystal clear. Tolerance to the venous and arterial effects of nitrates was found to occur at varying degrees, depending on the design of the specific study, and the arterial and venous parameters chosen for measurement. The relative contribution of arterial and/or venous tolerance to the ultimate loss of clinical efficacy remains to be defined.

Biochemical (Vascular) Mechanisms of Nitrate Tolerance

Metabolic activation to nitric oxide and vascular nitrate tolerance

It is now recognized that organic nitrates are metabolically converted to nitric oxide (NO) in vascular smooth muscle cells. The particular enzymes involved have not been identified, but they have been shown to be membrane bound [28] and not identical to glutathione-S-transferases [29]. Chromatographic evidence suggests that a common enzyme metabolizes organic nitrates to NO in the microsomal fraction of the vascular smooth muscle cell [30]. When vascular tolerance was produced, both relaxation and NO production from vascular microsomes were reduced [31]. These results, therefore, suggest that in vitro vascular nitrate tolerance is accompanied by reduced metabolic activation of organic nitrates to its pharmacologically active product, NO.

The reason for this reduced vascular metabolism has been postulated some years ago by Needleman and Johnson [32] in the form of the "sulfhydryl depletion hypothesis." This hypothesis states that "organic nitrates react with reduced sulfhydryl groups in the vascular smooth muscle receptor, leading to the formation of a disulfide linkage. [Tolerance is brought about by conversion of] the nitrate receptor to the disulfide form, which presumably has a lower affinity for NTG." This hypothesis suggests, therefore, that critical sulfhydryl groups in the vascular smooth muscle cells are involved in bringing about nitrate action. It is uncertain whether these critical sulfhydryl groups reside in the target "receptor" [32] or in the nitrate-metabolizing enzyme, or whether they simply serve as cofactors for the conversion process to NO.

In a series of elegant studies, Needleman and his associates have presented a substantial case for the involvement of critical sulfhydryl groups in promoting the in vitro vasorelaxant action of nitrates [33]. Although several recent studies are not in direct agreement with this hypothesis [34], the overwhelming evidence is in accord with this theory. There is little doubt that in isolated blood vessels exposed to high concentrations of nitrates some form of the sulfhydryl depletion hypothesis is operative in producing in vitro vascular tolerance toward nitrates.

Sulfhydryl depletion hypothesis of nitrate tolerance: How valid is it in vivo?

The validity of this mechanism as it applies to in vivo nitrate tolerance, however, is far from conclusively demonstrated. The conditions for creating nitrate tolerance in isolated vessel preparations (i.e., millimolar concentrations of NTG, exposure time generally less than 1 hour, lack of systemic regulation) are quite different from those obtained under clinical conditions (i.e., nanomolar concentrations of NTG, exposure time in hours, presence of systemic regulation), and it is therefore difficult (and perhaps unadvisable) to extrapolate conclusions obtained from in vitro preparations to in vivo situations.

The primary evidence that the sulfhydryl depletion hypothesis might be operative in vivo comes from the fact that sulfhydryl donors such as N-acetylcysteine (NAC) and methionine can reverse, or partially reverse, nitrate tolerance. Several studies have demonstrated this phenomenon in patients (Table 2). For example, Packer et al. [6] showed that partial reversal of hemodynamic tolerance toward intravenous NTG (at 6.4 μg/kg/min for 48 hours) can be obtained with oral NAC (200 mg/kg) in eight patients with congestive heart failure. May et al. [35] used intravenous NAC at 100 mg/kg for 15 minutes in seven patients who had received intravenous NTG at 45 ± 13 μg/min for 24 hours and showed that the intracoronary NTG dose-response curve in coronary sinus blood flow was restored with the sulfhydryl donor. Recently, Boesgaard et al. [36] showed that NAC can partially prevent (i.e., decrease the magnitude of) nitrate tolerance toward intravenous ISDN (at 5 mg/kg) in 10 patients with stable angina. The NAC dose used was 2 g over 15 minutes, then 5 mg/kg/hr for 30 hours, i.e., a total dose of 12.5 g of NAC for a 70 kg person). The clinical parameters monitored were increase in time to angina, increase in time to 1-mm ST depression, and reduction in ST-segment depression at comparable workload. These investigators showed that although the effects of ISDN declined with time, the rate of such decline was slower with NAC infusion compared to placebo.

Several other clinical studies showed a mixture of positive and negative results. Dupuis et al. [37] administered an intravenous NAC dose of 225 mg/kg/day to 13 patients with congestive heart failure who were on intravenous NTG at 1.5 μg/kg/min and showed that NAC did not prevent tolerance development in mean arterial pressure, ventricular filling pressures, and systemic vascular resistance. However, the right atrial pressure remained depressed with NAC infusion. Ghio et al. [23] gave a smaller dose of intravenous NAC, 5 g over 15 minutes, 24 hours after intravenous NTG (169 ± 115 μg/min) in 20 patients with coronary artery disease and showed that NAC could reverse venous tolerance (as measured by venous volume), but did not exert much effect on arterial parameters. Recently, Boesgaard et al. [38] examined the effect of simultaneous administration of NAC and ISDN on the development of nitrate tolerance in seven patients with stable angina pectoris. The treatments involved 8-day periods with ISDN (40 mg qid) together with NAC (controlled release 600 mg qid) or placebo. Exercise time and time to angina were unaffected by NAC, which, however, augmented the antiischemic effects of ISDN as assessed by ECG, albeit only partially.

Table 2. *Does sulfhydryl donation avoid nitrate tolerance in vivo?*

Study	Nitrate dose	Sulfhydryl	Population	Study design	Parameters	Results	Benefit?
Single dose sulfhydryl before and after nitrate tolerance development							
Levy et al. [41]	NTG td, 10 mg/day, 3 days	MET IV, 5 g	11 normal subjects	0.4 mg sl NTG challenge before/after MET	Forearm venous volume	MET potentiated NTG at baseline, restored sl NTG response	Yes
Ghio et al. [23]	NGT IV, 169 ± 115 μg/min, 24 hr	NAC IV, 5 g	20 patients with CAD	NAC administered after 24 hr NTG infusion	Forearm venous volume, arterial BP	NAC reversed venous tolerance, but no effect on arterial parameters	Yes/no
May et al. [35]	NTG IV, 45 ± 13 μg/min, 24 hr	NAC IV, 100 mg/kg	7 patients with CAD given NAC, 5 placebo	Intracoronary NTG dose-response before/after NAC	Coronary sinus blood flow	Reversal of NTG tolerance	Yes
Packer et al. [6]	NTG IV, 6.4 μg/kg/min, 48 hr	NAC oral, 200 mg/kg	8 patients with severe CHF	Single dose NAC administered to NTG-tolerant pts.	PCWP, MRAP, HR, SVI, MAP, SVR	Partial reversal of NTG tolerance	Yes

Long-term sulfhydryl/nitrate coadministration							
Hogan et al. [40]	NTG td, 20 mg/day, 4 days	NAC oral, 200 mg tid, 4 days	7 normal subjects	R, DB, C-O	BP, HR at rest and exercise	No prevention of NTG tolerance	No
Parker et al. [39]	ISDN oral, 40 mg qid, 7–10 days	NAC IV, 100 mg/kg	12 patients with stable angina	R, DB, PC	Treadmill walking time	No reversal of NTG tolerance	No
Hogan et al. [41]	NTG td, 10 mg/day, 4 days	NCA oral, 400 mg (or placebo) tid, 4 days	8 patients with stable angina	R, DB, C-O	BP, HR, treadmill walking time	No prevention of NTG tolerance	No
Boesgaard et al. [36]	ISDN IV, 5 mg/hr, 30 hr	NAC IV, 2 g over 15 min, 5 mg/kg/hr, 30 hr	10 patients with stable angina	R, DB, C-O	Bicycle time to angina, ST segment depression	Partial prevention of ISDN tolerance	Yes
Boesgaard et al. [38]	ISDN oral, 40 mg qid, 8 days	NAC oral, 600 mg (or placebo) SR, qid, 8 days	7 patients with stable angina	R, DB, C-O	Bicycle time to angina, ST segment depression	Exercise time to angina not changed, partial benefit of NAC on ECG changes	Yes/no
Dupuis et al. [37]	NTG IV, 1.5 μg/kg/min, 24 hr	NAC IV, 225 mg/kg/day	13 patients with CHF	R, C-O	MAP, SVI, SVR, PCWP, RAP, PAP	NAC did not prevent tolerance development, except lower RAP	Yes/no

td, transdermal; IV, intravenous; sl, sublingual; NTG, nitroglycerin; ISDN, isosorbide dinitrate; NAC, N-acetyl cysteine; MET, methionine; CAD, coronary artery disease; CHF, congestive heart failure; R, randomized; DB, double blind; C-O, cross-over; PC, placebo controlled; PCWP, pulmonary capillary wedge pressure; MRAP, mean right atrial pressure; BP, blood pressure; HR, heart rate; SVI, stroke volume index; SVR, systemic vascular resistance; PAP, pulmonary artery pressure.

Several other clinical studies, however, showed no NAC effect on nitrate tolerance. The first report, that of Parker et al. [39], showed that an intravenous dose of NAC at 100 mg/kg for 15 minutes (the same dose used by May et al. [35] to demonstrate a positive NAC effect) did not reverse tolerance development in 12 patients with stable angina given oral ISDN qid for 7-10 days. Treadmill walking time was used as the efficacy parameter. Hogan et al., in two separate studies, showed that oral NAC did not prevent tolerance development from transdermal NTG either in seven normal subjects [40] or in eight patients with stable angina [41]. The oral NAC doses used 200 mg tid in the study involving normal subjects versus 400 mg tid in that involving angina patients. An analysis of the results of these clinical studies suggest two conclusions: (a) NAC may exert a positive effect on clinical nitrate tolerance, particularly when it was used in high doses. (b) This positive effect of tolerance reversal was often only partial in extent and, in many cases, not observed at all. These findings, therefore, indicate that NAC coadministration is not a clinically viable approach to combate nitrate tolerance.

Two other sulfhydryl-generating agents have also been used in an attempt to prevent or reverse the occurrence of nitrate tolerance. The L-cysteine precursor, methionine, was tested by Levy et al. [42] as a 5 g intravenous dose in 11 volunteers given transdermal NTG at 10 mg/day for 3 days. Sublingual NTG (0.4 mg) was employed as a challenge dose and venous volume was measured using forearm plethysmography. Results of this study showed that methionine potentiated venous volume at baseline and also restored venous volume response to sublingual NTG after transdermal dosing. Münzel et al. recently examined the mechanisms of interaction of methionine and NTG in intact dogs [43,44]. Their results indicate that vascular smooth muscle, unlike hepatocytes, cannot convert methionine to cysteine, and that the observed effect of methionine is likely due to hepatic, rather than vascular, conversion to cysteine. The beneficial interactive effect occurred primarily in the resistance vessel circulation, and not in large epicardial arteries.

The coadministration of a sulfhydryl ACE inhibitor such as captopril in nitrate therapy is theoretically attractive, because such a compound could potentially counter nitrate tolerance on two fronts, i.e., repletion of sulfhydryl groups in the vascular and inhibition of neurohormonal regulation resulting from activation of the renin-angiotensin system. Conflicting results, however, exist regarding the ability of captopril to prevent or reverse nitrate tolerance, and there are data suggesting that a nonsulfhydryl ACE inhibitor such as enalapril can have a positive effect on nitrate tolerance [45]. In vitro studies indicate that captopril is less effective than NAC in reacting with NTG in plasma to generate NO [46]. Thus any positive effect of captopril (at its clinical doses) on nitrate tolerance is likely to be more due to its ACE inhibitor activity than to its sulfhydryl function. Details of these studies will be discussed later.

The overall data suggest that sulfhydryl-donating agents can potentiate nitrate action, and to a certain extent, reverse in vivo nitrate tolerance. We have pointed out, however, that these sulfhydryl effects (even if they were considered unequivocally positive) are not, by themselves, sufficient evidence to validate the sulfhydryl depletion hypothesis as the operative mechanism of nitrate tolerance. The reason for this conclusion is that sulfhydryl agents can react *extracellularly* with NTG outside the vascular smooth muscle cell, producing S-nitrosothiols or NO in the process (Figure 1). These products of nitrate metabolism could then cause the potentiation and tolerance-reversing effects seen [47].

Physiological (Systemic) Mechanisms of Nitrate Tolerance

An alternative to the classical vascular mechanism of nitrate tolerance has appeared recently in the literature and has gained considerable support. This alternative mechanism suggests that during chronic nitrate administration stimulation of endogenous vasoconstrictive forces occurs, which leads to the development of pharmacologic tolerance. The mediator(s) of these constrictive forces is likely to be a neurohormonal factor, such as renin, angiotensins, catecholamines, etc. Several lines of evidence suggest the presence of reflex vasoconstriction during nitrate administration, and they are outlined below.

Withdrawal rebound as an evidence of physiological counter-regulation

Perhaps the most direct evidence for the presence of physiological counter-regulation during nitrate administration is the occurrence of rebound hemodynamic changes after abrupt NTG withdrawal. Historical data showed that after years of chronic exposure to NTG, some munitions workers developed severe ischemic events after retirement [48]. In two recent studies, Packer et al. [49] and Olivari et al. [7] showed that abrupt cessation of transdermal NTG in CHF patients produces rebound decreases in cardiac index, and increases in mean arterial pressure and in

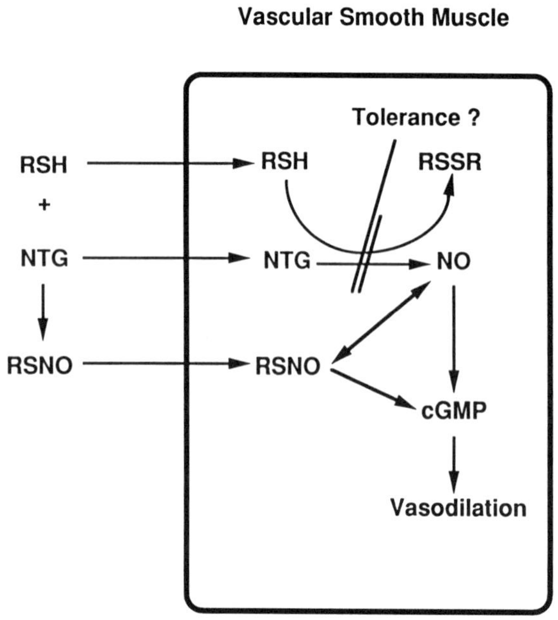

Fig. 1. *Schematic diagram of proposed intracellular and extracellular mechanisms by which thiols may potentiate the actions of NTG and reverse nitrate tolerance in the vascular smooth muscle cell. RSNO, S-nitrosothiol; RSH, thiol; RSSR, disulfide; NO, nitric oxide; cGMP, cyclic CMP. (Taken from Fung et al. [47], with permission.)*

systemic, pulmonary, and forearm vascular resistances. These rebound changes were transient and occurred within the first 2 hours after stopping the dose. In a recent study examining intermittent versus continuous NTG monotherapy for stable angina, Ferantini et al. [50] found that during a 25-day period a total of 11 night angina attacks occurred in 6 out of the 10 patients during the nitrate-free interval, and most of these occurred within the first 3 hours. These results are consistent with the theory that reflex vasoconstriction develops during drug administration, which is unopposed when NTG administration is stopped. These data cannot be readily explained by the sulfhydryl depletion hypothesis outlined earlier.

The factors that contribute to this withdrawal rebound phenomenon are not yet well defined, but probably relate to the pharmacokinetic properties of NTG [51]. It is possible that during nitrate withdrawal the short elimination half-life of NTG (in minutes) may remove the vasorelaxant effects so rapidly that the counter-regulatory vasoconstrictive effects, which are still operative, become unopposed. We tested this hypothesis in a rat model of CHF by comparing the rebound hemodynamics observed in animals whose NTG infusion was either stopped abruptly or through a graded withdrawal [51]. We showed that abrupt NTG withdrawal caused rapid and transient rebound increases in LVEDP. This elevation of LVEDP, above pre-NTG baseline values, peaked within 40 minutes after stopping the infusion and was completely avoided by a graded reduction of NTG infusion (20% reductions in infusion rate every 20 minutes). These results are consistent with the theory that the short biological half-life of NTG may play an important role in evoking hemodynamic rebound. The rebound phenomenon for NTG may be an important clinical problem in some patients, particularly when nitrates are used in monotherapy in an intermittent mode, where abrupt withdrawal occurs daily. It is possible that the occurrence of rebound would be less likely with the use of other nitrate formulations with longer half-lives.

Changes in plasma neurohormonal levels

Studies in normal volunteers have shown that continuous trandermal NTG therapy (average dose 29 mg/day) for 7 days led to a 50% reduction in urinary sodium excretion, decreases in hematocrit, and weight gain. In these same studies, plasma renin activity, and plasma concentrations of norepinephrine and arginine vasopressin were significantly elevated after the first 24 hours of NTG treatment [52]. Packer et al. [6] have also provided evidence for a role of neurohormonal activation in nitrate tolerance in CHF patients. Hemodynamic tolerance produced by continuous NTG infusion (6.4 μg/kg/min) for 48 hours was paralleled by increases in heart rate, plasma renin activity, and body weight. Dupuis et al. [37] also showed similar results in CHF patients in whom tolerance development during NTG infusion was accompanied by increases in plasma renin activity at 1 and 6 hours during infusion. In addition, rapid intravascular fluid shifts were observed during NTG infusion. Hematocrit levels were significantly reduced after the first hour and continued to drop throughout the 24-hour NTG infusion period. Corresponding increases in calculated total blood volume were also observed. Statistically significant increases in plasma renin activity and epinephrine were observed at after 1 hour of infusion, but no changes in arginine vasopressin, norepinephrine, or aldosterone were detected.

Other studies, however, did not observe significant increases in plasma neurohormones during nitrate administration. For example, Olivari et al. [7] found no elevation in plasma renin activity or norepinephrine levels during continuous transdermal NTG for 24 hours, while partial hemodynamic tolerance was observed. No change in heart rate was observed in the CHF patients during NTG, while normal subjects exhibited prompt elevations in heart rate and plasma norepinephrine levels during NTG. Therefore, the neurohormonal responses observed during nitrate administration (and possibly the tolerance processes) may be dependent upon the underlying disease state, and the nature and magnitude of the initial hemodynamic effects produced.

Changes in receptor density and affinity

Because of the high variabilities of the values, interpretation of changes in plasma levels of neurohormones during vasodilator therapy is often difficult and equivocal. In addition, most of these endogenous neurohormones act locally or regionally, and changes could occur at these local sites with little or no impact in the systemic plasma concentrations. Further, physiological counter-regulation can be brought about not only by changes in systemic neurohormonal concentrations, but by changes at receptor site as well (e.g., changes in receptor densities or affinities, or changes in receptor-effector coupling).

These points are illustrated by results of a recent study from our laboratory [53]. We showed that transdermal NTG administration to CHF rats for 10 days, either continuously or intermittently, caused a two-fold increase in isolated abdominal aorta responsiveness to phenylephrine in vitro. Significant increases in cardiac beta-adrenoceptors (as judged by [^{3}H]dihydroalprenolol binding) were also observed in the intermittent NTG treatment group when compared with controls. These results are consistent with the possibility that neurohormonal system changes may take place at levels other than systemic plasma concentrations of agonist. Nevertheless, it seems likely that neurohormonal alterations do play a significant role in limiting nitrate effectiveness during continuous administration, both in humans and experimental animals. Activation of the renin-angiotensin system and sodium and fluid retention may be particularly important, since these changes have been shown most consistently.

Pharmacokinetic-pharmacodynamic model of nitrate tolerance involving neurohormonal counter-regulation

Over the past several years we have demonstrated a rat model of CHF to be useful for the study of NTG-induced hemodynamic effects and in vivo tolerance development [54–56]. This animal model mimics the behavior of CHF patients with respect to initial hemodynamic effects of NTG infusion, the development of tolerance during continuous therapy, the reversibility of the tolerant state, and hemodynamic rebound after abrupt drug withdrawal [51,54]. We have recently developed a mathematical model in an attempt to describe the tolerance phenomenon and to gain insight into the various rate processes involved [57]. In this model we assumed that NTG produced a direct vasodilatory effect, which was unchanged during infusion, so long as the plasma concentration remained constant. We also assumed that counter-regulatory vasoconstrictive force was responsible for the development of NTG tolerance, and that this vasoconstriction was driven by the hypotensive effect of NTG administration. This model (Figure 2), therefore, was constructed to be consistent with the "neurohormonal activation" hypothesis of nitrate tolerance. Fitting such a mathematical model to our hemodynamic data in CHF rats provided an estimate of 2.2 hours for the first order half-life of disappearance of the counter-regulatory vasoconstrictive force. Thus, the proposed model predicts that after complete tolerance development during NTG infusion in these animals, a drug-free interval of 9–11 hours (i.e., 4–5 half-lives) would be required for these rats to regain complete responsiveness to NTG. This prediction is consistent with other studies we have carried out in this animal model [54] and also with clinical data [58]. This pharmacokinetic/pharmacodynamic model may therefore be useful for developing a better understanding of the rate processes governing tolerance development in vivo and for developing rational improvements in dosage regimen design for nitrates.

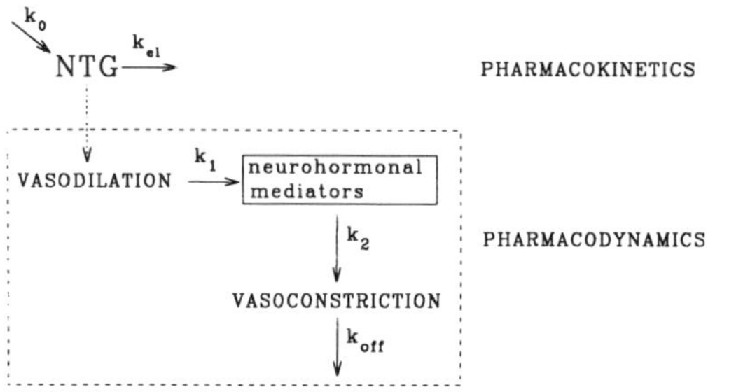

Fig. 2. *A pharmacokinetic-pharmacodynamic model of nitrate tolerance. k_0 is the drug input rate. k_{el} is the elimination rate constant of NTG. k_1 is the rate constant of production of counter-regulatory factors, which mediate a compensatory vasoconstriction with a rate constant k_2. k_{off} is the disappearance rate constant of this compensatory process. (Adapted from Bauer and Fung [57], with permission.)*

Why do nitrates evoke physiological counter-regulation?

The evidence presented indicates that development of nitrate tolerance is, at least in part, due to activation of neurohormonal systems. What are the causes of these compensatory responses? We have hypothesized that the predominant venodilating effect of nitrates may be an important factor involved in this phenomenon. In previous studies we compared the left ventricular hemodynamic effects and tolerance properties of an experimental S-nitrosothiol (S-nitroso N-acetylpenicillamine, SNAP) to those of NTG in our animal model [55]. We found SNAP to have a greater effect on afterload when compared to NTG, even at the same reductions in LVEDP, and also found no apparent hemodynamic tolerance development to SNAP, despite rapid tolerance to NTG during a 10-hour infusion. The result of no tolerance with SNAP is consistent with the sulfhydryl depletion hypothesis of nitrate tolerance, but the fact that SNAP had a different profile of hemodynamic action (i.e., a similar effect on preload, greater effects on afterload when compared to NTG) suggested that concomitant reductions in both preload and afterload may be an important determinant of long-term nitrovasodilator efficacy [55].

To further test our hypothesis that a "balanced" arterial and venous hemodynamic effect could maintain nitrate effectiveness, we coadministered hydralazine during NTG infusion in CHF rats [56]. We found that concurrent hydralazine prevented the hemodynamic tolerance observed during NTG infusion alone. Hydralazine itself produced no measurable reductions in preload when administered alone and had no effect on vascular NTG tolerance developed in vitro. These data are consistent with the hypothesis that the predominant mechanism of NTG-induced hemodynamic tolerance (at least during a 10-hour infusion) is neurohormonal rather than vascular. The beneficial interaction of hydralazine in the preload effects of NTG may explain the long-term clinical efficacy of hydralazine/ISDN combination in CHF [59], despite the fact that ISDN was administered in a dosage regimen that (alone) was likely to cause tolerance.

Why might concurrent arterial and venous action be important in preserving the preload effects of nitrates? We have previously speculated that renal perfusion may be an important factor involved in the nitrate tolerance process [56]. Leier et al. have shown that ISDN when administered alone causes decreases in renal blood flow (approximate reductions of 200 ml/min) and increases in renal vascular resistance in CHF patients [60]. Flaim also showed, in a rat model of CHF similar to ours, that NTG infusion produced approximately 20% reductions in renal blood flow [61]. It seems plausible, therefore, that nitrate-induced reductions in renal blood flow, or changes in renal blood flow distribution, may cause activation of renal neurohormonal factors (including increased renin release and sodium and fluid retention) in an attempt to preserve renal perfusion. In the same patient population mentioned above, Leier et al. found that hydralazine, when administered alone or in combination with ISDN, increased renal blood flow by about 100 ml/min and decreased renal vascular resistance [60]. This effect on renal perfusion was probably secondary to increases in cardiac output, since hepatic and limb blood flows were also increased. The effects of nitrates on renal perfusion in angina patients have not been examined. Further studies examining the role of renal perfusion on nitrate hemodynamic effects and tolerance are currently being carried out in our laboratory.

Can combination therapy avoid nitrate tolerance?

Aside from sulfhydryl donors, ACE inhibitors and diuretics have been examined for their potential benefit in avoiding nitrate tolerance. A summary of these studies is presented in Table 3. Dakak et al. [62] examined the effect of oral captopril (60 ± 25 mg/day) on the effects of intravenous NTG in patients with heart failure and showed no difference in nitrate effectiveness between the two groups. On the other hand, Mehra et al. [63] showed that the effects of oral ISDN (40–120 mg, q6h) in 14 CHF patients were more sustained when used in conjunction with oral captopril, indicating that tolerance toward ISDN might have been lessened in the presence of captopril. Using normal subjects and measuring forearm venous volume, Katz et al. [45] found that tolerance toward transdermal NTG can be avoided by both captopril and enalapril, suggesting that a sulfhydryl group is not required in the ACE inhibitor to produce this beneficial interaction.

Varriale et al. [64] showed that in CHF patients who did not respond to intravenous NTG, administration of furosemide (IV, 40 mg q6h) overcame their nitrate resistance. On the other hand, Parker et al. [65] found that administration of a combination of hydrochlorothiazide (50 mg) and amiloride (5 mg) did not prevent development of hemodynamic tolerance toward transdermal NTG in normal subjects.

These results showed that ACE inhibitors and diuretics may have a beneficial effect on the develop-

Table 3. *Can combination therapy avoid nitrate tolerance in vivo?*

Study	Nitrate dose	Interacting treatment	Population	Study design	Parameters	Results
ACE inhibitors						
Dakak et al. [62]	NTG IV, 465 ± 481 µg/min, 24 hr	Captopril oral, 60 ± 29 mg/day, before and during study	21 patients with severe CHF (12 NTG alone, 9 NTG + captopril)	2 parallel groups, NTG IV alone or with oral captopril	RAP, PCWP, CI, SAP, PVR, PRA, ALD, BW	No difference in NTG effectiveness between the 2 groups
Mehra et al. [63]	ISDN oral, 40–120 mg, q6h	Captopril oral, 89 ± 32 mg/day	14 patients with severe CHF	Captopril > 72 hr, then ISDN added to therapy	RAP, PAP, PAWP, SBP	ISDN effects more sustained, less tolerance with captopril
Katz et al. [44]	NTG td, 10 mg/day, 72 hr	Enalapril (10 mg bid, 7 days), or captopril (25 mg tid, 7 days), or placebo	34 normal subjects	sl NTG challenge at 2 hr and 72 hr after 7 days pretreatment with ACE I or placebo	Forearm venous volume	Tolerance only found in the placebo group
Diruetics						
Varriable et al. [64]	NTG IV, up to 400 µg/min	Furosemide IV, 40 mg q6h to nitrate nonresponders	26 patients with severe CHF	Nonresponders were given diuretic to decrease RAP to 10 mmHg, then NTG rechallenge	RAP, PCWP	NTG resistance was overcome after diuresis
Parker et al. [65]	NTG td, 19 ± 1 mg/day, 5–7 days	Hydrochlorothiazide/ amiloride (50 mg/5 mg), or placebo, 7 days	22 normal subjects	R, DB, PC	SBP, HR, HcT	No differences between diuretic and placebo groups

td, transdermal; IV, intravenous; sl, sublingual; NTG, nitroglycerin; ISDN, isosorbide dinitrate; NAC, N-acetyl cysteine; MET, methionine; CAD, coronary artery disease; CHF, congestive heart failure; R, randomized; DB, double blind; C-O, cross-over; PC, placebo controlled; PCWP, pulmonary capillary wedge pressure; MRAP, mean right atrial pressure; BP, blood pressure; HR, heart rate: SVI, stroke volume index; SVR, systemic vascular resistance; PAP, pulmonary artery pressure.

ment of clinical nitrate tolerance, particularly in selected patients with CHF. These data also indirectly argue for a significant role of neurohormonal compensation as a mechanism of in vivo nitrate tolerance, since some of the interacting compounds (e.g., enalapril and furosemide) are not expected to interfere with the sulfhydryl status and cellular metabolism of nitrates in the vascular smooth muscle cell. An inherent difficulty in designing and interpreting experiments involving these combination therapies is that both ACE inhibitors and diuretics can have significant hemodynamic effects on their own, and suitable controls must be incorporated before meaningful conclusions can be drawn.

Mechanisms of Nitrate Tolerance: Single or Multiple?

Based on current information it is possible to conclude that clinical nitrate tolerance may be brought about by more than a single mechanism. These mechanisms may be evoked at different stages after chronic nitrate administration, and their relative contributions may change depending on the physiologic or pathologic states of the patient. For example, in congestive heart failure, soon after nitrate dosing is commenced, neurohormonal counter-regulation may be initiated in an attempt to maintain renal blood flow. Continued nitrate administration may then bring about biochemical changes in the blood vessels, either in the form of sulfhydryl depletion or in the upregulation of vascular constrictive tone through other receptor-effector systems. Further studies are needed to define the relative extents and time courses of these mechanisms in regulating nitrate action and tolerance.

Acknowledgment

This work was supported in part by grants HL22273 and GM 42850 from the National Institutes of Health, USA.

References

1. Ahlner J, Andersson RGG, Torfgard K, Axelsson KL. Organic nitrate esters: Clinical use and mechanisms of actions. *Pharmacol Rev* 1991;43:351–423.
2. Elkayam U. Tolerance to organic nitrates: Evidence, mechanisms, clinical relevance, and strategies for prevention. *Ann Intern Med* 1991;114:667–677.
3. Cowan JC. Avoiding nitrate tolerance. *Br J Clin Pharmacol* 1992;34:96–101.
4. Kowaluk E, Hough K, Fung H-L. Effect of intermittent exposure and drug-free intervals on the in vitro vascular tolerance to nitroglycerin. *Life Sci* 1989;44:1157–1163.

5. Elkayam U, Kulick D, McIntosh N, Roth A, Hsueh W, Rahimtoola SH. Incidence of early tolerance to hemodynamic effects of continuous infusion of nitroglycerin in patients with coronary artery disease and heart failure. *Circulation* 1987;76:577–584.
6. Packer M, Lee WH, Kessler PD, Gottlieb SS, Medina N, Yushak M. Prevention and reversal of nitrate tolerance in patients with congestive heart failure. *N Engl J Med* 1987;317:799–804.
7. Olivari MT, Carlyle PF, Levine B, Cohn JN. Hemodynamic and hormonal response to transdermal nitroglycerin in normal subjects and in patients with congestive heart failure. *J Am Coll Cardiol* 1983;2:872–878.
8. Jordan RA, Seth L, Casebolt P, Hayes MJ, Wilen MM, Franciosa J. Rapidly developing tolerance to transdermal nitroglycerin in congestive heart failure. *Ann Intern Med* 1986;104:295–298.
9. Elkayam U, Jamison M, Roth A, Kulick D, Vasquez J, Rahimtoola SH. Oral isosorbide dinitrate in chronic heart failure: Tolerance development to QID vs. TID regimen [abstr]. *J Am Coll Cardiol* 1989;13:178.
10. Zimrin D, Reichek N, Bogin KT, et al. Antianginal effects of intravenous nitroglycerin over 24 hours. *Circulation* 1988;77:1376–1384.
11. Parker JO, Fung H-L. Transdermal nitroglycerin in angina pectoris. *Am J Cardiol* 1984;54:471–476.
12. Reicheck N, Priest C, Zimrin D, Chandler T, St. John Sutton M. Antianginal effects of nitroglycerin patches. *Am J Cardiol* 1984;54:1–7.
13. Poorvin D, Cabot CF, Chi PY, et al. Acute and chronic antianginal efficacy of continuous twenty-four-hour application of transdermal nitroglycerin. *Am J Cardiol* 1991;68:1263–1273.
14. Thadani U, Fung H-L, Darke AC, Parker JO. Oral isosorbide dinitrate in the treatment of angina pectoris. Dose response relationship and duration of action during acute therapy. *Circulation* 1980;62:491–502.
15. Thadani U, Fung H-L, Darke AC, Parker JO. Oral isosorbide dinitrate in angina pectoris. Comparison of duration of action and dose response relationship during acute and sustained therapy. *Am J Cardiol* 1982;49:411–419.
16. Parker JO, VanKoughnett KA, Fung H-L. Transdermal isosorbide dinitrate in angina pectoris. Effect of acute and sustained therapy. *Am J Cardiol* 1984;54:8–13.
17. Thadani U, Hamilton SF, Olson E, et al. Duration of effects and tolerance of slow-release isosorbide-5-mononitrate for angina pectoris. *Am J Cardiol* 1987;59:756–762.
18. Curfman GD, Heinsimer JA, Lozner EC, Fung H-L. Intravenous nitroglycerin in the treatment of spontaneous angina pectoris: A prospective, randomized trial. *Circulation* 1983;67:276–282.
19. Horowitz JD, Henry CA, Syrjanen ML, et al. Combined use of nitroglycerin and N-acetylcysteine in the management of unstable angina pectoris. *Circulation* 1988;77:787–794.
20. Diodati J, Theroux P, Latour J-G, Lacoste L, Lam JYT, Waters D. Effects of nitroglycerin at therapeutic doses on platelet aggregation in unstable angina pectoris and acute myocardial infarction. *Am J Cardiol* 1990;66:683–688.
21. Zelis R, Mason DT. Isosorbide dinitrate: Effect on the vasodilator response to nitroglycerin. *JAMA* 1975;234:166–170.
22. Makhoul N, Dakak N, Flugelman MY, et al. Nitrate tolerance in heart failure—differential venous, pulmonary and systemic arterial effects. *Am J Cardiol* 1990;65:J28–J32.
23. Ghio S, Deservi S, Perotti R, Eleuteri E, Montemartine C, Specchia G. Different susceptibility to the development of

nitroglycerin tolerance in the arterial and venous circulation in humans—effects of N-acetylcysteine administration. *Circulation* 1992;86:798–802.
24. Manyari DE, Smith ER, Spragg J. Isosorbide dinitrate and glyceryl trinitrate: Demonstration of cross tolerance in the capacitance vessels. *Am J Cardiol* 1985;55:927–931.
25. Leier CV, Huss P, Magorien RD, Unverferth DV. Improved exercise capacity and differing arterial and venous tolerance during chronic isosorbide dinitrate therapy for congestive heart failure. *Circulation* 1983;67:817–822.
26. Stewart DJ, Elsner D, Sommer O, Holtz J, Bassenge E. Altered spectrum of nitroglycerin action in long-term treatment: Nitroglycerin-specific venous tolerance with maintenance of arterial vasodepressor potency. *Circulation* 1986;74:573–582.
27. Goldberg RK, Lee RW, Olafos M, Goldman S. Development of tolerance to nitroglycerin in the arterial and venous circulation of dogs. *J Am Coll Cardiol* 1987;10:1335–1341.
28. Chung S-J, Fung H-L. Identification of the subcellular site for nitroglycerin metabolism to nitric oxide in bovine coronary smooth muscle cells. *J Pharmacol Exp Ther* 1990;253:614–619.
29. Chung S-J, Chong S, Seth P, Jung CY, Fung H-L. Conversion of nitroglycerin to nitric oxide in microsomes of the bovine coronary artery smooth muscle is not primarily mediated by glutathione-S-transferases. *J Pharmacol Exp Ther* 1992;260:652–659.
30. Chung S-J, Fung H-L. A common enzyme may be responsible for the conversion of organic nitrates to nitric oxide in vascular microsomes. *Biochem Biophys Res Commun* 1992;185:932–937.
31. Chung S-J, Fung H-L. Relationship between nitroglycerin-induced vascular relaxation and nitric oxide production: Probes with inhibitors and tolerance development. *Biochem Pharmacol* 1993;45:157–163.
32. Needleman P, Johnson EM. Jr. Mechanisms of tolerance development to organic nitrates. *J Pharmacol Exp Ther* 1973;184:709–715.
33. Needleman P (ed). *Organic Nitrates*. New York: Springer-Verlag, 1975.
34. Fung H-L, Chong S, Kowaluk E. Mechanisms of nitrate action and vascular tolerance. *Eur Heart J* 1989;10(Suppl A):2–6.
35. May DC, Popma JJ, Black WH, et al. In vivo induction and reversal of nitroglycerin tolerance in human coronary arteries. *N Engl J Med* 1987;317:805–809.
36. Bosegaard S, Aldershvile J, Poulsen HE. Preventive administration of intravenous N-acetylcysteine and development of tolerance to isosorbide dinitrate in patients with angina pectoris. *Circulation* 1992;85:143–149.
37. Dupuis J, Lalonde G, Lemieux R, Rouleau JL. Tolerance to intravenous nitroglycerin in patients with congestive heart failure—Role of increased intravascular volume, neurohumoral activation and lack of prevention with N-acetylcysteine. *J Am Coll Cardiol* 1990;16:923–931.
38. Bosegaard S, Aldershvile J, Pedersen F, Pietersen A, Madsen JK, Grande P. Continuous oral N-acetylcysteine treatment and development of nitrate tolerance in patients with stable angina pectoris. *J Cardiovasc Pharmacol* 1991;17:889–893.
39. Parker JO, Farrell B, Lahey KA, Rose BF. Nitrate tolerance: The lack of effect of N-acetylcysteine. *Circulation* 1987;76:572–576.
40. Hogan JC, Lewis MJ, Henderson AH. N-acetylcysteine fails to attenuate hemodynamic tolerance to glyceryl trinitrate in healthy volunteers. *Br J Clin Pharmacol* 1989;28:421–427.

41. Hogan JC, Lewis MJ, Henderson AH. Chronic administration of N-acetylcysteine fails to prevent nitrate tolerance in patients with stable angina pectoris. *Br J Clin Pharmacol* 1990; 30:573–577.
42. Levy WS, Katz RJ, Wasserman AG. Methionine restores the venodilative response to nitroglycerin after the development of tolerance. *J Am Coll Cardiol* 1991;17:474–479.
43. Münzel T, Mülsch A, Holtz J, Just H, Harrison DG, Bassenge E. Mechanisms of interaction between the sulfhydryl precursor L-methionine and glyceryl trinitrate. *Circulation* 1992;86: 995–1003.
44. Münzel T, Mülsch A, Harrison DG, Bassenge E. Mechanisms of interaction between the sulfhydryl precursor L-methionine and glyceryl trinitrate [abstr]. *Circulation* 1992;86:I487.
45. Katz RJ, Levy WS, Buff L, Wasserman AG. Prevention of nitrate tolerance with angiotensin converting enzyme inhibitors. *Circulation* 1991;83:1271–1277.
46. Chong S, Fung H-L. Biochemical and pharmacological interactions between nitroglycerin and thiols: Effects of thiol structure on nitric oxide generation and tolerance reversal. *Biochem Pharmacol* 1991;42:1433–1439.
47. Fung H-L, Chong S, Kowaluk E, Hough K, Kakemi M. Mechanisms for the pharmacologic interaction of organic nitrates with thiols. Existence of an extracellular pathway for the reversal of nitrate vascular tolerance by N-acetylcysteine. *J Pharmacol Exp Ther* 1988;245:524–530.
48. Lange RL, Reid MS, Tresch DD, Keelan MH, Bernhard VM, Coolidge G. Nonatheromatous ischemic heart disease following withdrawal from chronic industrial nitroglycerin exposure. *Circulation* 1972;46:666–678.
49. Packer M, Medina N, Yushak M, Lee WH. Hemodynamic factors limiting the response to transdermal nitroglycerin in congestive heart failure. *Am J Cardiol* 1986;57:260–267.
50. Ferantini M, Pirelli S, Merlina P, Silva P, Pollavini G. Intermittent transdermal nitroglycerin monotherapy in stable exercise-induced angina: A comparison with a continuous schedule. *Eur Heart J* 1989;10:998–1002.
51. Bauer JA, Fung H-L. Effect of apparent elimination half-life on nitroglycerin-induced hemodynamic rebound in experimental heart failure. *Pharm Res* 1993;10:1341–1345.
52. Parker JO, Parker JD. Neurohormonal activation during nitrate therapy: A possible mechanism for nitrate tolerance. *Am J Cardiol* 1992;70:93B–97B.
53. Bauer JA, Fung H-L, Kwon YW, Gopalakrishnan M, Triggle DJ. Continuous versus intermittent nitroglycerin administration in experimental heart failure: Vascular relaxation, adrenoceptor and ion channel effects. *J Cardiovasc Pharmacol* 1993;22: 600–608.
54. Bauer JA, Fung H-L. Effects of chronic glyceryl trinitrate on left ventricular hemodynamics in a rat model of congestive heart failure: Demonstration of a simple model for the study of in vivo nitrate tolerance. *Cardiovasc Res* 1990;24:198–203.
55. Bauer JA, Fung H-L. Differential hemodynamic effects and tolerance properties of nitroglycerin and an S-nitrosothiol in experimental heart failure. *J Pharmacol Exp Ther* 1991;256: 249–254.
56. Bauer JA, Fung H-L. Concurrent hydralazine administration prevents nitroglycerin-induced hemodynamic tolerance in experimental heart failure. *Circulation* 1991;84:35–39.
57. Bauer JA, Fung H-L. Pharmacodynamic models of nitroglycerin-induced hemodynamic tolerance in experimental heart failure. *Pharm Res* 1994;11:816–823.

58. Packer M, Gottlieb SS, Kessler PD, Medina N, Yushak M, Lee WH. What drug-free interval is required to prevent the development of organic nitrate tolerance in heart failure? *Circulation* 1981;76:IV255.
59. Cohn JN, Archibald DG, Zeische S, et al. Effect of vasodilator therapy on mortality in chronic congestive heart failure: Results of a Veterans Administration cooperative study (V-HeFT). *N Engl J Med* 1986;314:1547–1552.
60. Leier CV, Magorien RD, Desch CE, Thompson MJ, Unverferth DV. Hydralazine and isosorbide dinitrate: Comparative central and regional hemodynamic effects when administered alone or in combination. *Circulation* 1981;63:102–109.
61. Flaim SF. Peripheral vascular effects of nitroglycerin in a conscious rat model of heart failure. *Am J Physiol* 1982;243: H974–H981.
62. Dakak N, Makhoul N, Flugelman MY, et al. Failure of captopril to prevent nitrate tolerance in congestive heart failure secondary to coronary artery disease. *Am J Cardiol* 1990;66:608–613.
63. Mehra A, Ostrzega E, Shotan A, Johnson JV, Elkayam U. Persistent hemodynamic improvement with short-term nitrate therapy in patients with chronic congestive heart failure already treated with captopril. *Am J Cardiol* 1992;70:1310–1314.
64. Varriale P, David WJ, Chryssos BE. Hemodynamic resistance to intravenous nitroglycerin in severe congestive heart failure and restored response after diuresis. *Am J Cardiol* 1991; 68:1400–1402.
65. Parker JD, Farrell B, Fenton T, Parker JO. Effects of diuretic therapy on the development of tolerance during continuous therapy with nitroglycerin. *J Am Coll Cardiol* 1992;20:616–622.

Clinical Pharmacokinetics of Nitrates

Marc G. Bogaert

Summary. The pharmacokinetics of organic nitrates are discussed with emphasis on the possible clinical relevance. For glyceryl trinitrate, the measurement of plasma concentrations is very difficult. Its pharmacokinetics are unusual, with a very rapid disappearance from plasma, and large intraindividual and interindividual variations. After oral administration, there seems to be a very extensive first-pass hepatic extraction and the plasma concentrations are often below the detection limit; after sublingual administration, glyceryl trinitrate appears in plasma. With transdermal glyceryl trinitrate controlled-release systems, plasma concentrations of glyceryl trinitrate can be maintained over 24 hours, although with fluctuations and important intraindividual and interindividual variability. After administration of glyceryl trinitrate via different routes, glyceryl dinitrates and mononitrates are present in plasma. The pharmacokinetics of isosorbide dinitrate are somewhat easier to understand. The substance disappears less rapidly from the plasma than does glyceryl trinitrate. After oral administration, there is also a hepatic first-pass extraction; the plasma concentrations can be prolonged by administering slow-release products. Sublingual administration leads to higher plasma concentrations than oral administration. Isosorbide dinitrate is metabolized in the organism to isosorbide 5-mononitrate and isosorbide 2-mononitrate, which both have vasodilator activity: after administration of isosorbide dinitrate, the mononitrates, and mainly the 5-mononitrate, reach very high concentrations in plasma. Isosorbide 5-mononitrate has been studied in its own right as an antianginal agent: it is completely absorbed after oral administration; it has a half-life of around 4 hours, and oral standard and controlled-release formulations have been extensively studied. For several reasons, including attenuation of the effects with time and the appearance of active metabolites, the relationship between the plasma concentrations of nitrates and their therapeutic effects is very complex. Knowledge of the pharmacokinetics of these substances is only of limited interest.

Key Words. nitrates, glyceryl trinitrate, glyceryl dinitrates and mononitrates, isosorbide dinitrate, isosorbide 5-mononitrate, pharmacokinetics, plasma concentrations

Twenty years ago we published the first method for the measurement of the plasma concentrations of glyceryl trinitrate and of isosorbide dinitrate after administration of therapeutic doses of these substances in humans [1,2]. The purpose of this review is to summarize what we have learned about the pharmacokinetics of the nitrates in the last 20 years and also to discuss the possible clinical relevance of these data. The discussion will be limited to glyceryl trinitrate and its active metabolites (glyceryl dinitrates and mononitrates) and to isosorbide dinitrate and its active metabolites (isosorbide 2-mononitrate, but mainly isosorbide 5-mononitrate). Other nitrates, such as pentaerythroltetranitrate, will not be discussed, as there are hardly any recent data available. Emphasis will be on findings that could possibly help in answering the clinician's questions. Full details and references to older studies can be found in extensive recent reviews [3–7].

Glyceryl Trinitrate and Metabolites

Glyceryl trinitrate (or nitroglycerin) is metabolized in the organism, inter alia, to glyceryl dinitrates and glyceryl mononitrates, which have some vasodilator activity.

Methodologic considerations

In most studies, glyceryl trinitrate in plasma has been measured with a gas chromatographic method with electron capture detection (GC-ECD). Since we introduced this method [1,2], there have been many attempts to improve its accuracy and sensitivity. Gas chromatography—mass spectrometry (GC-MS) has also been used with success in some studies. Details of these methods can be found elsewhere [3,4,8].

Determination of the very low plasma concentrations of glyceryl trinitrate (from 0.1 ng/ml plasma upwards) is still difficult, and some published results should be viewed with caution. Apart from the assay itself, there are also problems related to the rapid breakdown of glyceryl trinitrate in blood and even in plasma; and the substance can be absorbed by or adsorbed to materials, necessitating an elaborate sampling procedure and precautions in storing and handling the samples before and during the assay. Moreover, concentrations can differ according to the sampling site. A possible influence of posture and exercise will be discussed later in this review. Some gas chromatographic assays allow the simultaneous determination of the parent compound and its metabolites [9,10].

Kinetics of glyceryl trinitrate after intravenous administration

In pharmacokinetic studies and also in clinical practice, attention should be paid to the loss of glyceryl trinitrate en route from the solution to the patient, in containers, infusion tubing, membrane filters, and catheters used for administration [11–13]. Adsorption is mainly important for PVC plasticizers and can be minimized by using glass containers, methacrylate butadiene styrene burettes, and polybutadiene tubing. For syringes, glass or polypropylene are appropriate.

Details of the different studies performed have been summarized [3,4,7]. For the same infusion rate, the plasma concentrations vary markedly from individual to individual but also within a given individual. The clearance values reported are much higher (often between 10 and 80 l/min) than the hepatic blood flow. Clearance by the liver (i.e., by hepatic biotransformation of glyceryl trinitrate to its metabolites) can therefore only explain part of the elimination. Glyceryl trinitrate is not excreted in unchanged form by the kidneys, and there is evidence that its extrahepatic metabolism occurs by clearance of the substance in the vasculature, a mechanism that has been extensively studied and discussed by the group of Fung [14,15]. The values published for the disappearance half-life of glyceryl trinitrate from plasma are in the range of a few minutes.

Oral administration

The results obtained with oral administration of nitroglycerin preparations—mostly controlled-release preparations—are widely divergent [3,4,7]. In more recent studies using adequate methodology, with different preparations of glyceryl trinitrate given orally, the concentrations were often below the detection limit ($\approx$0.1 ng/ml), and only at some sampling times was glyceryl trinitrate detected [16–19]. These studies seem to confirm that in humans as well there is a very extensive hepatic first-pass extraction after oral administration. The pharmacodynamic and therapeutic effects of glyceryl trinitrate after oral administration are possibly linked to the metabolites.

Sublingual and buccal administration

For different preparations and for different doses, peak plasma concentrations of about a few nanograms per milliliter were found [4]. These concentrations decrease rapidly once the sublingual preparation is removed, with a half-life of a few minutes, which is in the range of that found after intravenous administration. Pharmaceutical and patient-related factors (such as dryness of the mouth and administration technique) could influence absorption, but there are few systematic studies that would allow conclusions about these factors. In a systematic study of sublingual versus intravenous administration, an average bioavailability of 36% was found [21]; even in this well-controlled study in volunteers, there was a very large interindividual variability. There has also been interest in the use of glyceryl trinitrate in a sublingual spray. Comparative studies between these sprays and sublingual tablets are not available.

Transdermal administration

Transdermal administration of glyceryl trinitrate is of potential interest as it avoids the nearly complete first-pass hepatic extraction occurring after oral administration. Furthermore, continuous absorption could lead to maintenance of rather steady plasma concentrations over time, which, in view of the very short half-life of the drug, was thought to be advantageous.

The first transdermal glyceryl trinitrate preparations used were ointments. Details of these studies have been summarized elsewhere [4]. In order to obtain stable concentrations over the time of application, transdermal controlled-delivery systems were developed. Details of these devices and the studies can be found elsewhere [3,4,24]. Glyceryl trinitrate is released from such systems at a constant rate for at least 24 hours. All glyceryl trinitrate that is released is assumed to be absorbed, so that skin characteristics are thought to be no longer a source of variability. These systems are usually labeled according to the amount released in vitro over 24 hours (e.g., 5, 10, or 15 mg). For a given system,, this corresponds to the size of the patch. The amount released per 24 hours should not be confused with the amount of glyceryl trinitrate present initially within the patch, which is always higher.

Assuming more or less constant release from the transdermal system, one can expect rather constant plasma concentrations over the period of constant release, that is, 24 hours. In fact, in the first few hours plasma concentrations increase gradually until a plateau is reached, the very short half-life of the drug notwithstanding; one assumes that this is related to a depot and metabolism function of the skin [25]. Once the plateau levels are reached, the glyceryl trinitrate concentrations within each individual are far from constant, and this could be related to fluctuations in the release from the patch or, more likely, to fluctuations in the transdermal bioavailability. There are no satisfactory explanations for the large interindividual variations. There are no suggestions that the site of application, for example arm versus the chest, could be responsible. Lack of standardization of the study conditions could be a factor (see below for influence of posture and exercise). In one study in which interindividual variability was marked and unexpectedly high concentrations were seen, the authors suggested a possible role for the presence of congestive heart failure [26].

If the patch is removed after 24 hours and is immediately replaced by another patch, concentrations are reasonably well maintained. After removal of the patch without replacement, concentrations fall somewhat less rapidly than after stopping an intravenous infusion, which again points to a possible depot function of the skin.

Numerous comparative studies have been done to evaluate the performance of the different systems, for example, with regard to the plasma concentrations obtained, to the fluctuations within one patient, and to differences between patients. Perhaps due to the very large intrasubject and intersubject variability, in many studies no difference between different systems was found, and contradictory results have been reported. From the older data and from more recent publications [27–29], one can infer that if there are differences in performance between different systems they are probably smaller than the background interindividual and intraindividual variability and most likely are not clinically relevant.

As there is evidence that maintaining constant plasma concentrations of glyceryl trinitrate over 24 hours leads to marked attenuation of the hemodynamic and therapeutic effects (see elsewhere in this series), transdermal delivery systems with time-dependent release have been developed, for example, systems with rapid initial release of glyceryl trinitrate and a decrease of release with time [24,30].

Glyceryl dinitrates and mononitrates

The metabolites of glyceryl trinitrate have vasodilator properties and could contribute to the pharmacodynamic and therapeutic effects when the parent compound is given, as mentioned above for the oral administration of glyceryl trinitrate. High concentrations of these metabolites are found after administration of the parent compound by different routes [17–19,25,31,32].

In some publications the kinetics of the glyceryl dinitrates or mononitrates were studied after administration of these compounds [33–35]. These results concerning metabolites are of interest, but at this time do not help in answering the question as to how these metabolites participate in the pharmacodynamic of therapeutic effects seen after administration of the parent compound.

Factors of variability

The large interindividual and intraindividual differences and the discrepancies between the studies discussed above are often not explained. In several studies the influence of posture was evaluated. After sublingual administration, lower plasma concentrations were reported during sitting than during lying down [38]. During transdermal application, there was no influence of standing [39]. In our study with transdermal administration, plasma concentrations, on average, almost doubled in the sitting position as compared with the supine position [40].

Exercise has been the subject of several studies. Several groups found increased concentrations during exercise after patch application [41,42] and during intravenous infusion [42]. After patch application on starting exercise, we found a small decrease in plasma concentrations but afterwards very marked increases, with a maximum at 5 minutes postexercise; plasma concentrations were still increased 1 hour after stopping exercise [40]. The influence of exercise was reported to be more important after renewal of a patch [43]. There has been interest in temperature as a possible factor of variability after application of a transdermal system [41–44]. From the results concerning posture, exercise, and temperature, one can certainly conclude that standardization of study conditions is mandatory.

Although one can expect that hemodynamic changes, for example cardiac failure, would affect glyceryl trinitrate kinetics, no systematic studies have been performed. It has been suggested, however, that the high plasma concentrations found in some studies could be due to the presence of congestive heart failure.

Isosorbide Dinitrate and Metabolites

Isosorbide dinitrate is metabolized in the body to two nitrate products, which also have vasodilator activity: isosorbide 2-mononitrate is more vasoactive than isosorbide 5-mononitrate. In many studies, isosorbide dinitrate was administered, and the plasma concentrations of unchanged product, and often those of the metabolites, were measured. Isosorbide 5-mononitrate was also studied extensively in its own right, in view of its use as an antianginal agent.

Methodologic considerations

Since our description of the original method for measuring plasma concentrations of isosorbide dinitrate and its mononitrate metabolites in plasma after administration of therapeutic doses of the substance in humans [2], numerous modifications of the chromatographic method have been published [4,8]. The methods for determination of isosorbide nitrates present less difficulties than those for glyceryl trinitrate. Isosorbide dinitrate and, especially, the mononitrates are more stable than glyceryl trinitrate in plasma, and the choice of sampling site and sampling material is less critical.

Kinetics of isosorbide dinitrate after intravenous administration

In older publications, and also more recently [45], possible problems with loss of isosorbide dinitrate to the material used for intravenous administration have been studied. The precautions mentioned for glyceryl trinitrate also apply here. For isosorbide 5-mononitrate, loss to the infusion material does not present a problem [46].

Details of a number of intravenous studies, usually with infusion, can be found elsewhere [4,6,7]. Clearance values between 2 and 4l/min have been reported. These values are considerably lower than those reported for glyceryl trinitrate, but still higher than the hepatic blood flow, which also points to a nonhepatic route of elimination. In most studies, a biphasic decline of the plasma concentrations was found after stopping the infusion: the half-life of the β-phase ranged between 18 and 139 minutes.

Oral administration

There have been numerous studies of oral administration of isosorbide dinitrate, both in standard and controlled-release formulations [4,6,7]. With standard formulations the peak isosorbide dinitrate concentration appears rapidly, between 30 and 60 minutes after administration, and the concentrations decline afterwards with a half-life of around 30–50 minutes. In a recent publication a much longer half-life was reported, and the authors assumed that the shorter half-lives usually reported result from too short sampling times [47]. Bioavailability figures around 30% are found; it is assumed that incomplete bioavailability is due to first-pass hepatic extraction. Differences in absorption between the preparations used are possible.

In numerous studies plasma concentrations were measured after administration of controlled-release preparations, in comparison with standard formulations or with other controlled-release formulations. With some preparations it is possible to obtain more or less constant plasma concentrations of isosorbide dinitrate over several hours. In most studies the bioavailability calculated for slow-release preparations was somewhat lower than that for a standard formulation.

Several authors have found that after repeated oral administration of isosorbide dinitrate, the plasma concentrations were higher than after the first administration. Isosorbide dinitrate concentrations after oral administration were much higher in vascular and muscle tissue than in plasma [48].

Sublingual administration

In many studies it was observed that plasma concentrations of isosorbide dinitrate peak to values several times higher than those after standard oral formulations. Bioavailability is thought to be on average 40–60%, but much lower and much higher values have been reported; sublingual bioavailability is thought to be approximately twice that after oral administration of a standard formulation. The half-life seen after sublingual administration is similar to that seen after oral administration [49].

Transdermal administration

Transdermal administration has been studied much less for isosorbide dinitrate than that for glyceryl trinitrate. From some older and some recent studies [52,53], one can conclude that the substance is absorbed through the skin and that high plasma concentrations can be obtained after dermal ointment, patch, or spray.

Metabolites

In the first studies of isosorbide dinitrate administration it was observed that isosorbide 2-mononitrate and isosorbide 5-mononitrate were present in plasma. The peak plasma concentrations of the mononitrated metabolites are much higher than those of the parent compound in all studies, and the 5-mononitrate reaches higher concentrations than isosorbide 2-mononitrate.

From those studies pharmacokinetic parameters have been obtained for these metabolites; these metabolites, and mainly isosorbide 5-mononitrate, have also been administered as such, with direct evaluation of their pharmacokinetic parameters. In the studies in which isosorbide dinitrate was administered, the ratios between the concentrations of the metabolites and those of the parent compound, and those between the metabolites, differed according to the mode of administration (single vs. chronic), the route of administration, etc. When the metabolites themselves were administered orally and/or intravenously, it was found that the bioavailability after oral administration was nearly 100% and the volume of distribution was about 50 l. The elimination half-life of isosorbide 5-mononitrate is somewhat longer than 4 hours, and for isosorbide 2-mononitrate it is about 2 hours. The pharmacokinetics of these metabolites are, in comparison with those of isosorbide dinitrate, predictable and show very limited interindividual and intraindividual variability.

Numerous studies have appeared about the plasma concentrations after oral administration of different commercial preparations, standard or controlled-release, of isosorbide 5-mononitrate. With the controlled-release preparations, the peak concentrations are lower and decline less rapidly than with the standard formulation given in equivalent doses. In comparison with standard formulations, the controlled-release preparations show a slight loss of bioavailability.

Factors of variability

There was no difference between the plasma concentrations of isosorbide dinitrate between females and males after oral administration [52]. One group reported results suggesting circadian variation in the pharmacokinetics of isosorbide 5-mononitrate [53,54]. There are, as for glyceryl trinitrate, only speculations about a possible influence of aging on the kinetics of the isosorbide dinitrates [36,37]. We found no influence of exercise on isosorbide dinitrate kinetics [55].

Food retarded the appearance of isosorbide dinitrate in plasma after oral administration, but it did not seem to alter total bioavailability; for sustained-release products enhancement of bioavailability with a high fat meal was described. Food was found to influence the absorption rate of a slow-release isosorbide 5-mononitrate, again without altering total bioavailability [56].

There is no influence of renal failure on the kinetics of isosorbide dinitrate or of isosorbide 5-mononitrate, as shown by our group [57] and afterwards by many others; there is no marked influence of hemodialysis or hemoperfusion on isosorbide dinitrate or isosorbide 5-mononitrate concentrations [58,59]. No influence of cirrhosis on the kinetics of isosorbide 5-mononitrate was found, while a slower elimination rate was found in some patients for isosorbide dinitrate. A slight influence of the presence of ischemic heart failure on the absorption rate of isosorbide 5-mononitrate was described, as was a slight lowering of the elimination rate constant [60].

In several studies, no influence of beta-blocker administration on the plasma concentrations of isosorbide dinitrate was found [61–63]. In a recent study no pharmacokinetic interaction was seen between isosorbide 5-mononitrate and the epinine precursor ibopamine [64].

Clinical Relevance

When the assay methods for nitrates in plasma became available 20 years ago, it was hoped that knowledge of the disposition and kinetics of the substances would help to solve a number of clinical questions. This hope implied that there would be a reasonably good relationship between the plasma concentrations of nitrates and their therapeutic effects. Even then, there were already some reasons to doubt such a relationship, and by 1970 I wrote in a review about our work on nitrates in animals, "... These results obtained *in vivo* and *in vitro* are compatible with the idea that for nitroglycerin-induced vasodilatation, a high concentration of the substance is not sufficient, but that rather a sudden increase in nitroglycerin concentration is needed" [65].

In numerous publications, for example, those cited in this review, discussion about this relationship can be found. In a review in 1988 about the relationship between the pharmacokinetics and pharmacodynamics of the organic nitrates, the authors concluded their summary by stating "On the basis of available data, plasma concentrations of various nitrates do not reliably predict clinical effects" [5]. There is at this time no reasons to doubt this statement, with the following possible explanations. First, the hemodynamic and therapeutic effects of nitrates are attenuated with time. The mechanism of this attenuation (counter-regulation?, true vascular tolerance?) is still unknown, but it is certainly not linked to pharmacokinetics: plasma concentrations after chronic treatment are, if anything, higher than those at the start of treatment. It is paradoxical that the development of controlled-release transdermal systems was stimulated by the idea that prolonging the plasma concentrations could lead to a prolonged effect, while it is now accepted that attenuation occurs much more readily when steady concentrations of the nitrates are maintained. The discussion of a nitrate-free period, aimed at avoiding attenuation, will be found elsewhere: the duration of the nitrate-free period depends on the rate of disappearance of the vasoactive nitrates from the organism.

The search for a relationship between concentrations and effects is also complicated by the formation of vasoactive metabolites from the parent molecules. Attempts have been made to calculate, on the basis of the vasodilator potency of different nitrates, how much each nitrate-unchanged product or metabolite-could contribute to certain effects. Such calculations are very difficult and probably not very useful, as the attenuation also occurs with the metabolites.

Attempts have also been made to look specifically at the concentration-effect relationship at the beginning of nitrate administration, when attenuation is not yet apparent and metabolite formation—at least for some routes of administration—is not pronounced. Even then, it is difficult to find a meaningful concentration-effect relationship. One reason is probably that the hemodynamic and therapeutic effects of vasodilator drugs result from effects in various segments of the circulation, with reflex interactions. In these circumstances, one is unlikely to find a simple concentration-effect relationship. Moreover, there are experimental data suggesting a different sensitivity towards nitrates in different parts of the circulation (e.g., arterial versus venous); for glyceryl trinitrate this is made still more complex by arterial-venous concentration gradients. In addition, the possibility exists that some therapeutic effects of the nitrates depend, in part, on interactions with other systems, for example, platelets. There is also the fundamental problem that the effect on vessels could be related to concentration, rate of uptake, or rate of transformation of the nitrate within vascular cells.

References

1. Bogaert MG, Rosseel MT. Plasma levels in man of nitroglycerin after buccal administration. *J Pharm Pharmacol* 1972; 24:737–738.
2. Rosseel MT, Bogaert MG. GLC determination of nitroglycerin and isosorbide dinitrate in human plasma. *J Pharmaceut Sci* 1973;62:754–758.
3. Bogaert MG. Clinical pharmacokinetics of glyceryl trinitrate following the use of systemic and topical preparations. *Clin Pharmacokin* 1987;12:1–11.
4. Taylor T, Taylor IW, Chasseaud LF, et al. Pharmacokinetics and metabolism of organic nitrate vasodilators. *Prog Drug Metab* 1987;10:207–336.
5. Thadani U, Whitsett T. Relationship of pharmacokinetic and pharmacodynamic properties of the organic nitrates. *Clin Pharmacokin* 1988;15:32–43.
6. Schaumann W. Pharmacokinetics of isosorbide dinitrate and isosorbide-5-mononitrate. *Int J Clin Pharmacol Ther Toxicol* 1989;27:445–453.
7. Ahlner J, Andersson RGG, Torfgård K, et al. Organic nitrate esters: Clinical use and mechanisms of actions. *Pharmacol Rev* 1991;43:351–423.

8. Jaeger H, Lutz D, Michaelis K, et al. Determination of nitrates in plasma. *Drugs* 1987;33(Suppl 4):9–22.
9. Han C, Gumbleton M, Lau DTW, et al. Improved gas chromatographic assay for the simultaneous determination of nitroglycerin and its mono- and dinitrate metabolites. *J Chromatogr Biomed Appl* 1992;579:237–235.
10. Jørgensen M, Andersen MP. Simultaneous determination of nitroglycerin and its dinitrate metabolites by capillary gas chromatography with electron-capture detection. *J Chromatogr Biomed Appl* 1992;577:167–170.
11. De Rudder D, Remon JP, Neyt EN. The absorption of nitroglycerin by infusion sets. *J Pharm Pharmacol* 1987;39:556–558.
12. Hansen HC, Spillum A. Loss of nitroglycerin during passage through two different infusion sets. *Acta Pharm Nord* 1991;3:131–136.
13. Haas CE, Nowak DR, Mabb WA. Effect of using a standard polyvinyl chloride intravenous infusion set on patient response to nitroglycerin. *Am J Hosp Pharm* 1992;49:1135–1137.
14. Fung H-L. Pharmacokinetics and pharmacodynamics of organic nitrates. *Am J Cardiol* 1987;60:4H–9H.
15. Fung H-L, Blei A, Chong S. Interpretation of nitrate plasma concentrations. Effect of cardiac output on nitroglycerin pharmacokinetics in experimental animals. *Eur Heart J* 1988; 9(Suppl A):39–43.
16. Noonan PK, Benet LZ. The bioavailability of oral nitroglycerin. *J Pharmaceut Sci* 1986;75:241–243.
17. Yu DK, Williams RL, Benet LZ, et al. Pharmacokinetics of nitroglycerin and metabolites in humans following oral dosing. *Biopharm Drug Disp* 1988;9:557–565.
18. Nakashima E, Rigod JF, Lin ET, et al. Pharmacokinetics of nitroglycerin and its dinitrate metabolites over a thirtyfold range of oral doses. *Clin Pharmacol Ther* 1990;47:592–598.
19. Kwon H-R, Green P, Curry SH. Pharmacokinetics of nitroglycerin and its metabolites after administration of sustained-release tablets. *Biopharm Drug Disp* 1992;13:141–152.
20. Mettle JKL, Glew RC, Jones MC, et al. Glyceryl trinitrate absorption: Chewing versus sublingual administration. *Br J Clin Pract* 1986;40:376–380.
21. Noonan PK, Benet LZ. Incomplete and delayed bioavailability of sublingual nitroglycerin. *Am J Cardiol* 1985;55:184–187.
22. Sanders SW, Michaelis K, Maurette JM, et al. Relative bioavailability of two spray formulations of nitroglycerin. *J Pharmaceut Sci* 1986;75:244–246.
23. Huber T, Merz P-G, Harder S, et al. Bioequivalence of sublingual glyceryl trinitrate. *Arzneim Forsch/Drug Res* 1990;40:1319–1322.
24. Wolff H-M, Bonn R. Principles of transdermal nitroglycerin administration. *Eur Heart J* 1989;10(Suppl A):26–29.
25. Nakashima E, Noonan PK, Benet LZ. Transdermal bioavailability and first-pass skin metabolism: A preliminary evaluation with nitroglycerin. *J Pharmacokinet Biopharm* 1987;15:423–437.
26. Jordan RA, Seth L, Casebolt P, et al. Rapidly developing tolerance to transdermal nitroglycerin in congestive heart failure. *Ann Intern Med* 1986;104:295–298.
27. McAllister A, Mosberg H, Settlage JA, et al. Plasma levels of nitroglycerin generated by three nitroglycerin patch preparations, Nitradisc, Transiderm-Nitro and Nitro-Dur and one ointment formulation, Nitrobid. *Br J Clin Pharmacol* 1986;21:365–369.
28. Noonan PK, Gonzalez MA, Ruggirello D, et al. Relative bioavailability of a new transdermal nitroglycerin delivery system. *J Pharmaceut Sci* 1986;75:688–691.

29. De Ponti F, Luca C, Pamparana F, et al. Bioavailability study of three transdermal nitroglycerin preparations in normal volunteers. *Curr Ther Res* 1989;46:111–120.
30. Wiegand A, Bonn R, Wagner F, et al. Haemodynamic effects of glyceryl trinitrate following repeated application of a transdermal delivery system with a phasic release profile. *Eur J Clin Pharmacol* 1991;41:115–118.
31. Noonan PK, Benet LZ. Variable glyceryl dinitrate formation as a function of route of nitroglycerin administration. *Clin Pharmacol Ther* 1987;42:273–277.
32. Laufen H, Leitold M. The pattern of glyceryl nitrates after oral administration of glyceryl trinitrate. *Arzneim Forsch/Drug Res* 1988;38:103–105.
33. Laufen H, Leitold M. Glyceryl-1-nitrate pharmacokinetics in healty volunteers. *Br J Clin Pharmacol* 1987;23:287–293.
34. Gumbleton M, Cashman JR, Benet LZ. 1,2- and 1,3-dinitrate metabolites of nitroglycerin: Spectroscopic characterization and initial administration to man. *Int J Pharm* 1991;71:175–186.
35. Leitold M, Laufen H, Yeates RA. Untersuchungen zur Pharmakologie and Pharmakokinetik von Glycerol-2-nitrat. *Arzneim Forsch/Drug Res* 1987;37:692–698.
36. Alpert JS. Nitrate therapy in the elderly. *Am J Cardiol* 1990;65:23J–27J.
37. Kelly JG, O'Malley K. Nitrates in the elderly. Pharmacological considerations. *Drugs and Aging* 1992;2:14–19.
38. Curry SH, Kwon H-R. Influence of posture on plasma nitroglycerin. *Br J Clin Pharmacol* 1985;19:403–404.
39. Heidemann R, Beckenbauer C, Woodcock BG. Effect of posture on glyceryl trinitrate plasma concentrations following transdermal application. *Br J Clin Pharmacol* 1987;23:246–247.
40. Lefebvre RA, Bogaert MG, Teirlynck O, et al. Influence of exercise on nitroglycerin plasma concentrations after transdermal application. *Br J Clin Pharmacol* 1990;30:292–296.
41. Barkve TF, Langseth-Manrique K, Bredesen JE, et al. Increased uptake of transdermal glyceryl trinitrate during physical exercise and during high ambient temperature. *Am Heart J* 1986;112:537–541.
42. Weber S, de Lauture D, Rey E, et al. The effects of moderate sustained exercise on the pharmacokinetics of nitroglycerin. *Br J Clin Pharmacol* 1987;23:103–105.
43. Gjesdal K, Klemsdal TO, Rykke EO, et al. Transdermal nitrate therapy: Bioavailability during exercise increases transiently after the daily change of patch. *Br J Clin Pharmacol* 1991;31:560–562.
44. Klemsdal TO, Gjesdal K, Bredesen J-E. Heating and cooling of the nitroglycerin patch application area modify the plasma level of nitroglycerin. *Eur J Clin Pharmacol* 1992;43:625–628.
45. De Muynck C, Colardyn F, Remon JP. Influence of intravenous administration set composition on the absorption of isosorbide dinitrate. *J Pharm Pharmacol* 1991;43:601–604.
46. De Muynck C, Colardyn F, Remon JP. The absorption of isosorbide-5-mononitrate to intravenous delivery systems. *J Pharm Pharmacol* 1990;42:433–434.
47. Smith D, Aldrich W, Dey M, et al. A pharmacokinetic model for isosorbide dinitrate, isosorbide-2-mononitrate, and isosorbide-5-mononitrate. *Drug Metab Disp* 1990;18:429–434.
48. Schneider W, Menke G, Heidemann R, et al. Concentrations of isosorbide dinitrate, isosorbide-2-mononitrate and isosorbide-5-mononitrate in human vascular and muscle tissue under

steady-state conditions. *Eur J Clin Pharmacol* 1990;38:145–147.

49. Keller-Stanislawski B, Marschner J-P, Rietbrock N. Pharmakokinetik von niedrigdosiertem Isosorbiddinitrat und Metaboliten nach bukkaler und oraler Gabe. *Arzneim Forsch/Drug Res* 1992;42:17–20.
50. Menke G, Schnellhammer R, Rietbrock N. Konzentrations-Zeit-Profil von Isosorbiddinitrat und seinen Metaboliten im Plasma nach perkutaner Resorption aus einem transdermalen therapeutischen System. *Arzneim Forsch/Drug Res* 1987;37:1301–1303.
51. Laufen H, Leitold M. Bioavailability and metabolism of isosorbide dinitrate from a transdermal spray. *Arzneim Forsch/Drug Res* 1992;42:931–935.
52. Nakatsu K, Brien JF, Savard G, et al. Plasma disposition and hemodynamic effects of a single oral dose of isosorbide dinitrate in human males and females. *Biopharm Drug Disp* 1992;13:357–367.
53. Lemmer B, Scheidel B, Blume H, et al. Clinical chronopharmacology of oral sustained-release isosorbide-5-mononitrate in healthy subjects. *Eur J Clin Pharmacol* 1991;40:71–75.
54. Scheidel B, Lemmer B. Chronopharmacology of oral nitrates in healthy subjects. *Chronobiol Int* 1991;8:409–419.
55. Bogaert MG, Lefebvre RA, Rosseel MT, et al. Influence of exercise on plasma concentrations of isosorbide dinitrate. *Eur J Clin Pharmacol* 1988;35:213–215.
56. Thomson AH, Miller SHK, Green ST, et al. The effect of food on the absorption of slow-release isosorbide-5-mononitrate tablets. *Eur J Clin Pharmacol* 1988;34:47–50.
57. Bogaert MG, Rosseel MT, Boelaert J, et al. Fate of isosorbide dinitrate and mononitrates in patients with renal failure. *Eur J Clin Pharmacol* 1981;21:73–76.
58. Bauer H, Laufen H, Franz HE. Isosorbide dinitrate in plasma and dialysate during haemodialysis. *Eur J Clin Pharmacol* 1986;30:187–190.
59. Evers J, Bonn R, Boertz A, et al. Pharmacokinetics of isosorbide-5-nitrate during haemodialysis and peritoneal dialysis. *Eur J Clin Pharmacol* 1987;32:503–505.
60. Meissner A, Petersenn S, Heidemann HT, et al. Pharmacokinetics of oral isosorbide-5-mononitrate in patients with ischemic heart failure. *Klin Wochenschr* 1991;69:213–219.
61. Chasseaud LF, Doyle E, Taylor T. Plasma concentrations and bioavailability of isosorbide dinitrate and pindolol from a combination formulation. *Biopharm Drug Disp* 1981;2:273–281.
62. Bogaert MG, Rosseel MT, Bruyneel K. Single administration of atenolol does not influence the kinetics of orally given isosorbide dinitrate. *Int J Clin Pharmacol Res* 1983;3:491–493.
63. Ochs HR, Neugebauer G, Greenblatt DJ, et al. Influence of beta-blocker coadministration on the kinetics of isosorbide mononitrate and dinitrate. *Klin Wochenschr* 1986;64:1213–1216.
64. Lefebvre RA, De Wilde G, Rosseel MT, et al. Investigation of a possible pharmacokinetic interaction between ibopamin and isosorbide-5-mononitrate. *Eur J Clin Pharmacol* 1992;42:549–552.
65. Bogaert MG, Herman AH, De Schaepdryver AF. Effects of nitroglycerin (trinitrin) on vascular smooth muscle. *Eur J Pharmacol* 1970;12:215–223.

Short and Long-Acting Oral Nitrates for Stable Angina Pectoris

Udho Thadani and Raymond J. Lipicky

Summary. **Nitroglycerin (NTG) spray and sublingual tablets rapidly relieve an established attack of angina, and their infrequent use is not associated with the development of tolerance. Although, following a suitable nitrate-free interval, the first dose of oral, long-acting nitrates produces significant hemodynamic effects, increases angina free walking, and decreases exercise-induced ischemia, during continued longterm therapy tolerance limits their usefulness. Appropriate dosing regimens of controlled-release formulations of isosorbide dinitrate (ISDN) and controlled-release NTG during long-term therapy have not been established. Use of immediate-release formulation of 15–120 mg of IDSN in a qid regimen lead to a marked reduction in the size and duration of antianginal effects compared to the initial dose. Asymmetric tid therapy with 30 mg of ISDN (7 a.m., 1 p.m, and 6 p.m.) is also associated with the development of partial tolerance and appears to provide antianginal prophylaxis for only a period of 6 hours each day. Asymmetric bid therapy with ISDN at 7 a.m. and noon may give sustained effect but is supported by only a single, small study that did not examine effectiveness after the noon dose in long-term use. Isosorbide-5-mononitrate (IS-5-MN) has been the subject of more recent studies than other nitrates because of attempts to bring a number of products into the U.S. market. IS-5-MN in qid, tid, and standard bid (8 a.m. and 8 p.m.) dosing regimens produce tolerance. Asymmetric regimens of immediate-release IS-5-MN (10 and 20 mg) given bid (once in the morning and again 7 hours later) decrease the development of tolerance compared to symmetric regimens and produce an increased exercise duration after each dose of the day; the 20 mg bid dosing is more effecive. Similarly, once-daily 120 and 240 mg controlled-release IS-5-MN does not produce tolerance and gives a sustained increase in daytime exercise duration. Both asymmetric bid immediate-release and once-daily controlled-release IS-5-MN preparations do not produce deterioration in exercise performance prior to the administration of the medication in the morning (i.e., no zero-hour effect). Further studies are needed to establish**

The opinions expressed here are those of the authors and should not be taken as those of FDA.

useful dosing regimens for ISDN, for controlled-release ISDN, and for controlled-release nitroglycerin. None of the dosing regimens of any oral, long-acting nitrate (including IS-5-MN) provide 24 hour antianginal and antiischemic effects. Although it is intuitively attractive to add an agent (beta-blocker or calcium channel blocker) that exerts 24 hour antianginal and antiischemic effects, which appear to depend on mechanisms different from those of nitrates, there are no studies that allow an evaluation of the usefulness of adding beta-blockers or calcium channel blockers to nitrate therapy (or vice versa).

Key Words. **stable angina pectoris, isosorbide dinitrate, exercise, isosorbide-5-mononitrate, oral nitroglycerin, buccal nitroglycerin, pentaerythritol tetranitrate, combination treatment**

Nitroglycerin (NTG) and other oral nitrates are extensively used for the treatment of patients with stable angina pectoris [1–3]. Since the first use of NTG in 1879 by Murrell [4], short and rapidly acting nitrates have become established as the treatment of choice to shorten the duration of pain of an established attack of angina or to prevent an anticipated attack by administering the agent just prior to physical activity or emotional stress [5–8]. The role of longer acting nitrates is much less well defined. When oral, long-acting nitrate therapy is first initiated or is withheld for a day or longer after chronic use, it can be shown [9–15] to increase angina free walking time and to prolong the time to exercise-induced myocardial ischemia, but continued use is associated with rapid development of tolerance [16–22]. Although knowledge of the appropriate doses or dosing intervals for many long-acting oral nitrates is only rudimentary, it is clear that the development of tolerance can be lessened [3] by choosing appropriate interdosing intervals, specifically by providing one "long nitrate" low or free interval determined by dose, the elimination half-life of the acive agent(s), and the release characteristics of the formulation. Use of an interdosing interval during a single 24-hour period does not provide continuous 24-hour antianginal effects from any oral nitrate formulation.

In this review, the pathophysiology of stable angina pectoris and the mechanism of action of nitrates is briefly discussed, and published data related to efficacy of various short and long-acting nitrate formulations are described. Conclusions are principally drawn from parallel-group, concurrent, placebo-controlled, exercise tolerance trials. This reflects our belief that baseline-controlled (i.e., lacking a concurrent placebo group) and, with some exceptions, crossover trials are difficult to interpret and can be misleading. Deficiencies in information are highlighted, and comments with respect to the appropriate use of various nitrate formulations are derived from evaluable data.

Role of Nitrates in Stable Angina Pectoris

There are no controlled clinical trials that have been designed to evaluate whether, compared to placebo, the use of long-acting oral nitrates has an effect on morbidity and/or mortality when used as a treatment of patients who have chronic stable angina pectoris. Oral nitrates are intended only to provide symptomatic relief (decreased angina frequency and prolongation of walking distance prior to development of angina, or relief of established angina despite continuing exercise). Their antiischemic properties (i.e., increase in walking distance to the development of significant ECG ischemia or wall motion abnormalities) are considered to be of value, although this seems less certain. This may be a moot point, however, as for nitrates the antianginal and antiischemic effects appear inseparable. When antianginal effects are discernable in studies, so are the antiischemic effects. When tolerance to the antianginal effects occur, so does tolerance to the antiischemic effects.

Pathophysiology of Stable Angina

The culprit lesion responsible for myocardial ischemia and angina pectoris is often the severe atherosclerotic eccentric lesion of one or more coronary arteries. This lesion limits segmental coronary blood flow during exercise or emotionally induced stress. The increase in myocardial oxygen requirement during stress cannot be met by a corresponding increase in segmental coronary blood flow, and this produces an imbalance between myocardial oxygen supply and demand, and leads to myocardial ischemia and angina pectoris. The endothelium of atherosclerotic vessels is dysfunctional [23–25] and stimuli that normally produce vasodilation, paradoxically produce vasoconstriction [23,24] due to deficiency of the endothelial derived relaxing factor [25–27]. The majority of patients with stable angina have reproducible exertion or excitement-induced angina. Many patients, however, also experience occasional episodes of angina at rest, probably due to a spontaneous increase in coronary vascular tone [28]. Only a minority of patients with stable angina do not have atherosclerotic heart disease, and in these patients reduction of coronary blood flow reserve probably plays a major role in the production of myocardial ischemia [3].

Mechanisms of Action of Nitrates in Stable Angina

Nitrates are effective smooth muscle relaxants and at low concentrations produce venodilation and changes in arterial compliance [29–32]. Arterial and arteriolar dilation occurs at higher concentrations [32]. Following nitrate administration, venous beds of the abdominal vasculature, arms, and legs become more compliant, blood volume in these regions is increased [30], and less blood returns to the heart and lungs. This results in a decrease in ventricular volume and pressure both at rest and during exercise [31]. Left ventricular wall stress is reduced with a resultant decrease in myocardial oxygen demand. A modest decrease in arterial pressure also reduces oxygen demand, but this is offset in part by a reflex increase in heart rate.

Nitrates produce dilation of coronary arteries when endothelium is denuded [26,33–37] or is dysfunctional [23,24,26]. Nitrates also dilate coronary stenosis [33] and increase collateral flow and improve subendocardial perfusion in the ischemic regions [1]. Nitrates relieve coronary spasm and prevent exercise-induced coronary constriction at the epicardial stenosis site [24]. Nitrates also reduce platelet aggregability and adhesions and thus may play a role in preventing coronary narrowing [38].

The current cellular model for these vascular effects is that after entering the smooth muscle cell, nitrates are converted to nitric oxide (NO) [27,33–36]. This process requires sulfhydryl groups [34]; NO or nitrosothiols stimulate soluble guanylate cyclase to produce cyclic guanosine monophosphate [37]. The cyclic nucleotide thus formed causes vasodilation as myocytic intracellular calcium is decreased, either by inhibition of calcium ion entry or by production of calcium exit [37]. Since nitrates are an exogenous source of NO, these agents are currently considered to replace the deficient NO in the vessels involved with atherosclerosis.

Short-Acting Nitrates and Single Doses of Long-Acting Nitrates

Nitroglycerin (NTG) administered as sublingual tablets or as an oral spray is rapidly absorbed and exerts its peak effect within 2–5 minutes [6–8,39]. These formulations of NTG are the mainstay for the rapid relief of an established attack of angina [5]. The tablets (0.4mg NTG) or spray (0.4mg spray) can be repeated every 5 minutes up to a maximum of three tablets or puffs. Maximum hemodynamic effects are observed in the upright posture, and patients should be advised to sit or stand rather than lie down in order to derive maximum benefit. Infrequent use of these agents is not associated with the development of tolerance. However, because of their short duration of action, these agents are not suitable for continuous or prolonged antianginal prophylaxis [1].

Although the sublingual form of isosorbide dinitrate (ISDN) in a dose of 5–15mg is used for the relief of an established attack of angina, it has a slower onset of action (peak effects in 5–10 minutes) than sublingual NTG or NTG spray, and for this reason sublingual NTG or NTG spray are the preferred agents [7].

Sublingual NTG tablets or spray and sublingual ISDN increase angina free walking time on the treadmill [6–8,39] for 1–2 hours and are suitable, when administered just prior to engaging in short-term physical or emotional stress, as prophylaxis against anticipated angina. Buccal NTG is effective for 3–5 hours [40–42] and is another formulation suitable for patients about to encounter short-term physical or emotional stress.

Although single doses of long-acting oral nitrates are not suitable for the relief of an established attack of angina, because of slow onset of action they can be used intermittently, like buccal NTG, to provide longer prophylaxis for up to 3–8 hours [9–15] in contemplation of exercise. So long as long periods of non-use follows each use, these agents can reliably prolong exercise tolerance for 3–8 hours, compared to placebo [9–15,43–47]. When frequency of use increases to more than once a day, however, tolerance can develop and the specific use pattern must be studied.

Use of Long-Acting Oral Nitrates Regularly

Isosorbide dinitrate (ISDN), and isosorbide-5-mononitrate (IS-5-MN), the major active metabolite of ISDN, available as immediate-release and controlled-release formulations, and controlled-release formulations of NTG (oral or topical) and an immediate-release pentaerythritol tetranitrate (PETN) are widely used in multiple-dose regimens for the long-term prophylaxis of angina pectoris. Recent data, however, show that the development of tolerance is a major limitation of the usefulness of such regimens, and that these regimens may or may not be superior to placebo [45–47].

A general review of the literature prior to 1973 revealed 38 studies in which effects of various nitrate preparations ISDN (n = 17), PETN (n = 14), and NTG (n = 7) were summarized [44]. Taking all studies together and simply taking the authors' conclusions at face value, 24 reported the drugs to be effective and 14 found them to be ineffective [44]. Of investigations that meet reasonable criteria for adequacy (double-blindness, randomization, use of placebo cotnrol, and a visible statistical analysis), only four found that oral, long-acting nitrates were efficacious [43]. In many of these studies, the doses of nitrates used were smaller than those that have subsequently been shown to exert hemodynamic and antianginal effects [43] in single doses, but more recent studies using appropriate doses have continued to yield variable results [45,46].

The importance of tolerance has only recently became generally recognized, and the variables that affect its development are only currently becoming defined. The development of several formulations of IS-5-MN for marketing in the United States (1993) occurred as the phenomenology of tolerance was becoming defined. Development of drugs was aided by recognition that a dosing regimen that avoided or reduced tolerance had to be found. All other oral, long-acting nitrate formulations were developed before tolerance was recognized as an important factor in their use, indeed before there was an effectiveness requirement in U.S. law (1962). Oral and topical (NTG ointment) long-acting nitrates (other than PETN) were effective, although as noted above, results of trials were variable. It is possible that their apparent persistent effect over time was related to the use of relatively low doses and relatively short half-lives or release periods. In addition, NTG ointments, for example, lose much of slow activity because of sublimation, so they may be, in effect, a nitrate-free interval even when they are used repeatedly. Real concern about tolerance grew markedly when NTG patches became available. These allow the round-the-clock exposure to significant NTG levels and induce rapid tolerance. It appears now that an appropriate regimen must be discovered for each dosage form of long-acting nitrates. Despite their widespread clinical use, and several recently conducted, well-designed trials, the appropriate use of many oral nitrate formulations cannot be readily deduced from available data, and the variables that affect the development of tolerance are still not well defined.

Isosorbide Dinitrate (ISDN) Immediate-Release Formulations

Studies by Thadani and colleagues [48] clearly documented effects, compared to placebo, of the first dose (after a washout period of 7 days) of 15, 30, 60, and 120 mg of ISDN on blood pressure and walking distance time as well as the attenuation but not complete loss of both the peak effect and duration of effect on each of these variables, despite greater plasma concentrations, after 1 week of regularly spaced qid daily therapy (Figure 1). These observations were confirmed by other investigators [49], and it has also been shown that the beneficial effects of the first dose of ISDN on exercise-induced ST-segment depression were similarly attenuated during long-term, regularly spaced tid therapy [50]. Although tolerance to the vasodepressor effects during ISDN therapy had been described previously [20,22], such effects were not considered relevant by many because of the prevailing notion that there was a

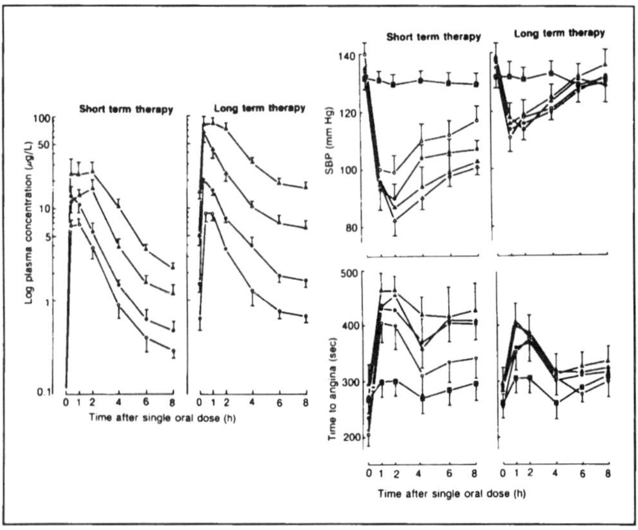

Fig. 1. *Plasma concentrations and duration of effects of isosorbide dinitrate (ISDN) during ocute and during sustained therapy qid in patients with angina pectoris. The 0-hour values during sustained therapy represent values 10 hours after a previous dose taken the night before. Plasma ISDN concentrations were higher during sustained therapy, but peak effects and duration of effects on systolic blood pressure at rest in the standing position, as well as exercise time to angina, were markedly attenuated. (Adapted from Thadani et al. [48], with permission.)*

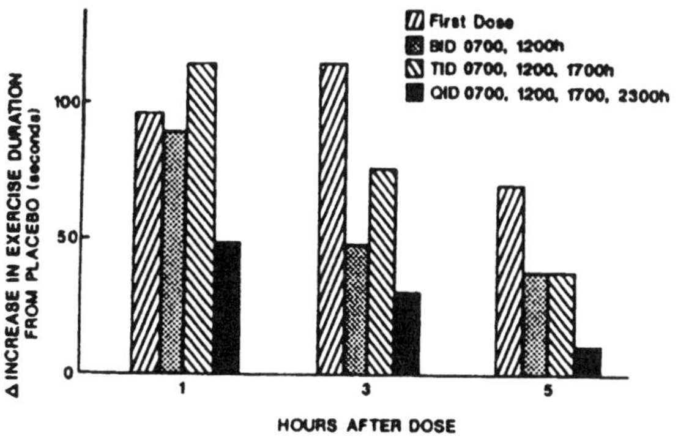

Fig. 2. Duration of effects after first dose of 30 mg of isosorbide dinitrate and after the same dose after 1 week of bid therapy given at 0700 and 1200 hours; tid therapy at 0700, 1200, and 1700 hours; and after qid therapy given at 0700, 1200, 1700, and 2300 hours. Data show a marked attenuation of effects after qid therapy compared with the first dose effect. Some attenuation was present, however, even after the less frequent regimens. (Adapted from Parker et al. [54], with permission.)

dissociation between circulatory and clinical tolerance [51,52].

Apparently contradictory data were published by Danahy and colleagues [51], who showed, with an average dose of 29 mg (range 20–50 mg) ISDN administered tid for 6 months, continued effects on exercise tolerance. In their study, however, nitrate therapy was withheld for 16–48 hours prior to the assessment of the effects of ISDN on exercise treadmill time during the long-term therapy evaluation [43], so that, in fact, it did not test a continuous regimen. Schneider and colleagues [52], in a baseline controlled trial, found decreased angina frequency during long-term therapy with ISDN 120 mg qid, but this study design is defective. Angina rates often decline in placebotreated patients in controlled trials [53].

Parker and colleagues, in a 12-patient study, reported persistent effectiveness with no tolerance with bid and tid dosing regimens of 30 mg ISDN [54]. Reanalysis of their data, however, showed attenuation of effects (Figure 2) compared to the effects of the first dose [43]. More important, the treatment regimens were asymmetric, providing a nitrate-free interval. In the bid regimen the drug was given at 7 a.m. and noon, and in the tid regimen drug was given at 7 a.m., noon, and 5 p.m. The effects of the next morning dose were evaluated after a time lapse of 15 and 20 hours after the previous last dose taken during the tid and bid regimens, respectively. Effects of the second and third dose of the day were not evalu-

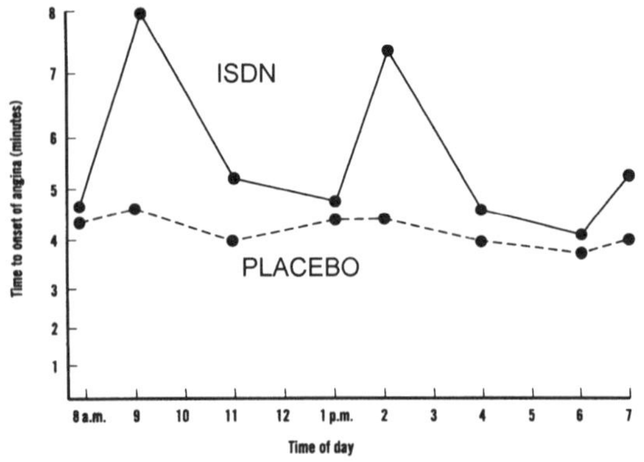

Fig. 3. *Day-long pattern of antianginal effects of isosorbide dinitrate. Mean duration of exercise until onset of angina at different hours after administration of isosorbide dinitrate (solid line) or placebo (dashed line). Capsules containing the active drug or placebo were given at 0800, 1300, and 1800 hours. A clinically significant increase in exercise time was present for only 1–2 hours after each dose. (Adapted from Bassan [55], with permission.)*

ated. Although the effects of the first dose of each day were not abolished due to the development of tolerance, the total duration of effect was not documented.

In a subsequent study conducted with 10 patients, Bassan [55] showed the development of tolerance even during asymmetric tid therapy (dosing at 8 a.m., 1 p.m., and 6 p.m.). In this study (Figure 3) it appeared that such a regimen increased exercise duration for a total period of only 6 hours out of the 9 hours of dosing, although the first dose of each day did produce a measurable effect and was consistent, as were the data of Parker and colleagues, with the notion that tolerance was not complete when there was a nitrate-free interval overnight. Also consistent with such a notion was an early report from Germany that showed that by using ISDN bid (in the morning and again 7 hours later), the development of tolerance could be attenuated [50]. Unfortunately this study measured only ST-segment depression during exercise and not exercise duration; the study also did not examine the effects of the afternoon dose.

It can be concluded that continued therapy with regularly spaced qid or tid regimens of immediate-release ISDN will result in the production of tolerance; in the case of qid, tolerance may be substantial and studies may not be able to differentiate ISDN in this regimen from placebo (e.g., Figure 2). Such regimens appear to be of no value. As for transdermal nitroglycerin (see review of ointment and patches) continued efficacy of a chronic regimen of ISDN depends upon introducing a daily nitrate-free interval; for immediate-release ISDN this interval appears to be at least 14–16 hours for a 30mg dose. There are no good data on the interval needed for other doses. Because the duration of effect of ISDN is likely to be principally due to its longer half-life active metabolite IS-5-MN, studies with IS-5-MN ought to be relevant to the use of ISDN, but further dose-ranging trials of immediate-release ISDN are also needed to define the necessary nitrate-free interval and to identify the range of useful dosing regimens.

Controlled-Release ISDN Formulations

Current labeling for controlled-release forms of ISDN call for tid or qid usage. The first dose of 40, 80, or 120mg doses of controlled-release ISDN have been shown to increase exercise duration for several hours [14,56]. Lee et al. reported beneficial effects of controlled-release ISDN 60mg qid compared to placebo during long-term therapy [57]. Silber reported development of tolerance during 80mg bid therapy but not when a dose of 160mg was given once a day or when bid dosing was given at 0800 and 1400 hours [56]. He did not, however, evaluate the effects of ISDN on exercise duration. In a subsequent large, multicenter study, a controlled-release formulation of 40mg ISDN, given bid at 0800 and 1400 hours, was not superior to placebo during long-term therapy (unpublished data presented at the June 2, 1993 Food and Drug Administration, Cardiac and Renal Drugs Advisory Committee [CRAC] Meeting).

At present it appears that regularly spaced qid, tid, and bid dosing regimens of controlled-release ISDN cannot be readily differentiated from placebo during chronic administration. Although single daily doses of these drugs are clearly effective, an effective way to administer more than one daily dose has not been defined. Consistent with the recommendations of the CRAC meeting, labelling for these formulations of ISDN as well as for all oral long-acting nitrates is being modified to reflect this uncertainty.

IS-5-MN Immediate-Release Formulations

IS-5-MN, the principal active metabolite of ISDN has a longer duration of action than its parent compound. It has no active metabolites and is nearly 100% bioavailable after oral administration [58]. In studies conducted in Europe, many of which were not double blind or placebo controlled, tid or qid therapy was reported to produce tolerance, while bid therapy with 20 or 40mg doses was said to show continued effectiveness [59–64]. Subsequent studies, however, showed that therapy with 20 and 40mg bid at 0800 and 2000 hours did produce tolerance after 1 week of therapy [65–66] and could not be considered an effective regimen.

The development of tolerance to IS-5-MN has been shown to be lessened when the drug was given in a dose of 20mg in the morning and 7 hours later for 1–2 weeks [67–71]. In one study, both morning and afternoon doses increased exercise duration compared to placebo with a total duration of effect of at least 12 hours (Figure 4) at the end of 1–2 weeks of therapy [70]. These findings have been confirmed with another formulation of IS-5-MN [72]. It is important to note that although the second dose of the day had some effect, it had a smaller effect than the first dose of the day, even though the plasma concentrations were higher [72]. Moreover, although the first dose of the day clearly lasted 7 hours, the effects of the second dose did not appear to last more than 4 hours [72]. Thus, although this asymmetric dosing regimen produces effects each day of a bid dose regimen and has an effect for more than 11 hours, tolerance dose occur during the course of a single day's treatment.

That some tolerance occurs wih this asymmetric (one dose in the morning and another 7 hours later) dosing regimen of immediate-release IS-5-MN is more

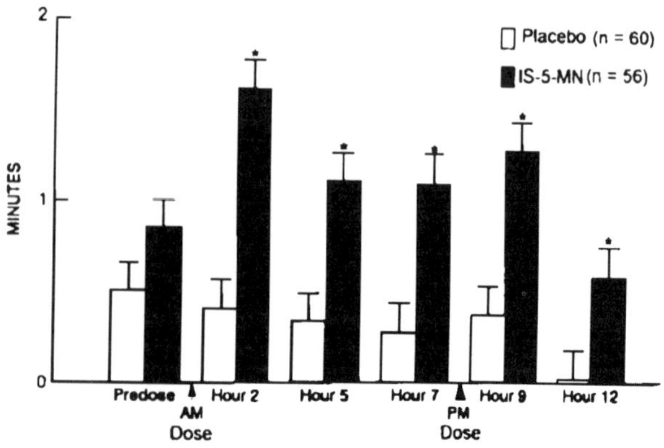

Fig. 4. *Changes in exercise duration during placebo and isosorbide-5-mononitrate. Mean changes in total exercise duration from respective baseline values after 0800 hour and 1500 hour doses of IS-5-MN and placebo. Total exercise duration increased significnatly ($p < 0.01$) with IS-5-MN compared to placebo at 2, 5 and 7 hours after the 0800 hour dose, and at 2 and 5 hours after the 1500 hour dose (hours 9 and 12 after the 0800 hour dose). Exercise time prior to the 0800 hour dose did not change significantly. Mean values ± SE are shown by vertical lines. *$p < 0.01$. (Adapted from Thadani et al. [70], with permission.)*

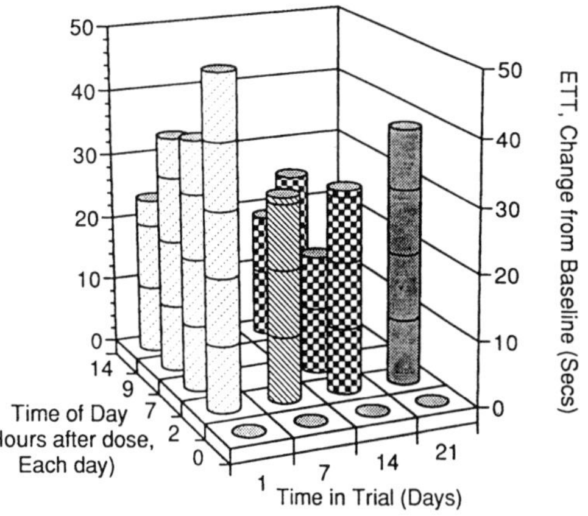

Fig. 5. *Mean (placebo-subtracted) results of symptom-limited exercise tolerance testing from a randomized, parallel-group, concurrent, placebo-controlled, multicenter trial that studied doses of 5, 10, and 20 mg of immediate-release isosorbide-5-mononitrate given in an asymmetric regimen. Dosing was in the morning just after the 0-hour exercise tolerance test and again at 1500 hours, just after the 7-hour exercise tolerance test. The mean results shown are for the 20 mg dose. Results for the 5 and 10 mg doses were qualitatively similar but of smaller magnitude. The hour 0 exercise tolerance measured on days 1, 7, 14, and 21 for the treated group was not statistically significantly different from placebo on any of the days and, for puposes of the figure, are arbitrarily represented as 0. Clearly, the greatest effect measured was with the first dose on the first day. All subsequently measured responses from the same dose were smaller. (Adapted from data on file with the Food and Drug Administration, with permission.)*

clearly shown in Figure 5 (unpublished data on file in the Food and Drug Administration, permission to use the data given by the anonymous owners). These are the placebo subtracted, mean results (change from baseline) of symptom-limited treadmill exercise tolerance times from a double-blind, concurrent, placebo-controlled, multicenter, dose-ranging (5, 10, and 20 mg), parallel-group trial that randomized a total of 165 patients [72]. Only the results from the 20 mg dose group are shown in Figure 5, but results with the 5 and 10 mg dose were similar. Doses were administered daily in the morning and again 7 hous later (in Figure 5, just after the 0 hour exercise tolerance test and then again 7 hours later just after measurement of exercise tolerance on days 1 and 14). The greatest effect measured was with the first dose on the first day, and all subsequently measured responses to the same dose were smaller.

None of the studies of asymmetrically administered immediate-release IS-5-MN showed rebound nocturnal angina or statistically significant deterioration in exercise duration prior to the morning dose during long-term therapy [70,72], as has been reported for intermittent therapy with nitroglycerin patches [73]. A dose of 20mg of IS-5-MN provides near-maximum improvement in exercise duration, and this improvement is only slightly attenuated when 20mg is used bid, once in the morning and again 7 hours later [70,72]. This regimen can give improved exercise performance for at least 14 hours. There remains no regimen that can produce round-the-clock exercise improvement.

IS-5-MN Controlled-Release Formulations

Many IS-5-MN controlled-release formulations are used in Europe. The results of studies are, however, conflicting. In one small study in seven patients, both 50 and 100mg doses were not superior to placebo at any time after 1 week of therapy, and no therapeutic efficacy could be documented at 24 hours after a single 100mg dose [74]. Other studies reported improvement in exercise duration for up to 8 hours after daily dosing with 60mg, with no development of tolerance [75–78]. The 60mg once-daily dose is the only dose used in Europe.

The results of a recently completed [79] double-blind, concurrent placebo-controlled, parallel-group, dose-ranging U.S. multicenter study that randomized 313 patients (270 completing and fulfilling all protocol criteria) are shown in Figure 6 (data on file at the FDA, permission has been granted to use the data). Measurements of exercise duration were made at 0, 4, and 12 hours after dosing on days 1, 14, 28, and 42 after randomization. Only the mean results (placebo subtracted) of the exercise tolerance test 12 hours after dosing are shown; 4 hour results were qualitatively similar, although the effects were greater in magnitude. As seems to be the case for all oral, long-acting nitrates, the greatest effect is seen with the first dose on the first day after a relatively long nitrate-free washout. Thereafter effects become smaller. This study had one unusual result. The 30 and 60mg doses had no effect on exercise duration by the 42nd day, indicating substantial tolerance, but doses of 120 and 240mg, once a day, continued to give a clear effect at 12 hours on day 42, suggesting that the higher doses could overcome some degree of tolerance without themselves provoking even greater tolerance. This is different from other experience.

It is also unclear why twice-daily asymmetric regimens of a total of only 20–40mg per day do not produce tolerance [70,72], while such doses given once daily in a slow-release formulation give substan-

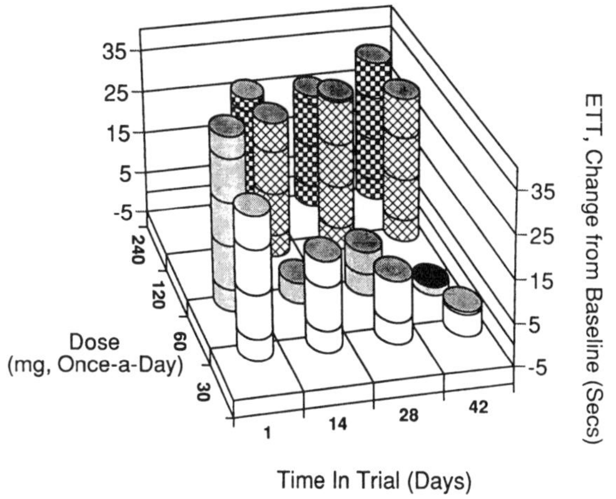

Fig. 6. *Mean (placebo-subtracted) results of symptom-limited exercise tolerance testing from a randomized (313 patients randomized, 270 completing and fulfilling all protocol criteria), parallel-group, concurrent, placebo-controlled, multicenter trial that studied doses of 30, 60, 120, and 240 mg of controlled-release isosorbide-5-mononitrate given once daily. Results shown were measured 12 hours after a morning dose, and measurements 4 hours after dose were qualitatively similar, although of greater magnitude. Note that at day 42, 12 hours after dosing the mean response obtained from the 60 mg group was 1.9 seconds less than that for the placebo group. Patients assigned to the 120 and 240 mg group were titrated to the assigned dose during the first week of the trial. On day 1 of the trial, patients assignd to 30 mg received 30 mg, patients assigned to 60 mg received 60 mg, and patients assigned to 120 or 240 mg were also receiving only 60 mg. Continued efficacy of the 120 and 240 mg doses, but not of the 60 mg dose, is apparent. (Adapted from data on file with the Food and Drug Administration, with permission.)*

tial tolerance. The transdermal nitroglycerin cooperative study [80] showed that long-term continuous high doses of trandermal nitroglycerin induced tolerance just as readily as with very low doses and that the tolerance could not be overcome by higher doses. Similarly, tolerance and crosstolerance to sublingual nitroglycerin develops within 24 hours of high-dose intravenous nitroglycerin therapy [81]. In any case, although not explained, it appears that 120 and 240 mg of a controlled-release formulation of IS-5-MN can provide about 12 hours of improved exercise tolerance when administered once daily.

Buccal Nitroglycerin

Buccal nitroglycerin 3 mg tid for 14 days has been shown to increase exercise duration, compared to placebo, and this effect was discernible at 5 hours after the morning dose [49]. This tid regimen was administered at 8 a.m., 1 p.m., and 5 p.m., and thus was a somewhat asymmetrical, not equally spaced tid regimen, as 15 hours elapsed between the last dose and the next morning dose, when the effects of treatment were assessed. Effects of the afternoon and evening doses on exercise were not evaluated. The efficacy of the second and third doses, therefore, remains unknown [43]. Although this sinlge study supports an asymmetric dosage regimen for this formulation, support for this regimen of this formulation must be considered weak, and further study is necessaery to fully support its use.

Nitroglycerin Tablets

Controlled-release formulations of NTG in doses of 2.5–18mg tid or qid are claimed to provide antianginal prophylaxis. In a double-blind, crossover study, 6.5mg slow-release NTG was shown to increase exercise time on the treadmill at 1 hour postdose after 2 weeks of theray, compared to placebo [82]. However, the authors of the study state that the increase in exercise duration in this sudy was less than they had previously observed after the first dose of this formulation of NTG [83]. A small but significant reduction in number of angina attacks was observed in the NTG tablet-treated patients compared to placebo [82]. However, the baseline frequency of angina was not documented. This study also did not evaluate any carryover effects. Previous studies reporting efficacy of controlled-release NTG did not use an adequate placebo group [84]. At present there are no adequate dose-response or placebo-controlled studies documenting effectivenes of controlled-release NTG tablets during long-term therapy in stable angina pectoris. Further study is required to define its appropriate use.

Pentaerythritol Tetranitrate (PETN)

Although still marketed with labeling that states PETN is possibly effective for the long-term prophylaxis of angina pectoris, review of available data through 1984 by the Food and Drug Administration led to the conclusion that there were no existing data supporting the effectiveness of PETN. *Federal Register* Notice (Volume 49, Number 200, page 40213, October 15, 1984) announced the opportunity for a hearing, and a hearing was held in 1989 (Docket Number 77N-0240; DESI1786). The administrative law judge was in favor of the Food and Drug Administration, recommending to the commissioner that PETN be removed from the market. The commissioner's final action has not yet taken place.

Can tolerance to oral nitrates be prevented by pharmacological means?

The exact cellular mechanism of tolerance is unknown. Depletion of SH groups, neuroendocrine activation, and increase in intravascular volume have all been implicated [45]. Studies designed to evaluate the prevention of tolerance by concomitant administration of SH group donors, ACE inhibitors, and diuretics have not produced convincing evidence [85–97] that tolerance can be prevented by pharmacological means [45].

Combination Therapy of Long-Acting Nitrates and Beta-Blockers or Calcium Channel Blockers

On theoretical grounds, a combination of a beta-blocker and a nitrate would be desirable, as they lower myocardial oxygen consumption by different mechanisms [45]. Studies designed to evaluate the effects of such combination therapy should be able to document that each of the agents in the combination contributes to the benefit (e.g., increased treadmill time) by showing that the combination is superior to each of the single ingredients and to placebo.

The majority of combination studies to date have been performed with nitrate dosing regimens that are now known to produce tolerance, and there are no studies in which effects of monotherapy with either agent alone have been compared to combination therapy in a systematic manner [98]. There is thus no documentation that oral, long-acting nitrates are useful in combination with other antianginal therapy. Short-acting nitrates, of course, are effective in the presence of other antianginal agents, and there is no reason to think that longer acting agents would not be.

In recent studies, patients who were symptomatic despite therapy with a beta-blocker have been studied during intermittent eccentric bid therapy with IS-5-MN [70,72,98]. Improvement in exercise duration was documented in this group of patients during IS-5-MN therapy compared to placebo, but it is unknown whether similar results would have been obtained with IS-5-MN monotherapy [98], that is, whether the beta-blocker had any effect at all in these patients. Similarly, in a recent study, patients who were symptomatic despite atenolol therapy improved after addition of IS-5-MN, and this combination was as effective as triple therapy with atenolol plus IS-5-MN and nifedipine [99]; it can be concluded that nifedipine did not contribute in that setting and that all three agents are tolerated together.

Adverse Effects of Long-Acting Nitrates

Headaches are the major side effect during therapy with long-acting nitrates, and dosing regimens that produce tolerance also lead to disappearance of headaches [43,80]. Intermittent dosing regimens that minimize or prevent the development of tolerance allow headaches to occur in some 30–50% of patients. Patients often learn to live with a mild headache, and nonprescription analgesics can produce relief [45]. Severe headaches (5–10% incidence) may occur in patients receiving intermittent nitrate therapy, necessitating withdrawal of therapy.

Nitrate-induced syncope can occur and is probably unreported [43]. Postural hypotension and symptoms of cerebral ischemia have been reported. If a patient is known to have experienced syncope after sublingual NTG, intermittent therapy with oral nitrates should be given with great care. Methemoglobin production can occur, but even with doses of nitroglycerin patches as great as 105 mg/24 hr for 30 days, methemoglobin plasma concentrations were not appreciable [80].

Conclusions

Although intermittent use of short-acting, oral nitrates (to terminate an attack of angina or to provide prophylaxis for anticipated activity) and intermittent use of long-acting oral nitrates (to provide prophylaxis for anticipated activity) provide useful antianginal therapy, tolerance limits multiple dosing regimens of oral long-acting nitrates and round-the-clock prophylaxis has not been achieved by any dosing regimen of any long-acting oral nitrate.

A dosing regimen for immediate-release IS-5-MN formulations for 20 mg (a dose in the morning and anotehr 7 hours later) has been shown to maintain antianginal and antiischemic effects for about 14 hours after the first dose of the day and should be a clinically useful formulation. A once-a-day, controlled-release formulation of IS-5-MN has been shown to maintain antianginal and antiischemic effects for about 12 hours after dosing at doses of 120 and 240 mg, once a day and should also be a clinically useful formulation. That a total daily dose of 40 mg (20 mg in the morning and again 7 hours later) of immediate-release IS-5-MN appear (from available data) to provide maximum effects with minimal tolerance, but smaller doses in a controlled-release form produce complete tolerance, while large doses of 120–240 mg controlled-release IS-5-MN do not is inexplicable at present.

With the exception of IS-5-MN, data are insufficient to show that multiple daily dosing regimens of other long-acting nitrates are effective long-term therapy in angina pectoris. Further studies are needed with immediate-release and controlled-release formulations of ISDN as well as with controlled-release NTG to establish multiple daily dosing regimens that are effective beyond the first dose during long-term therapy. For 30 mg of immediate-release ISDN, data weakly support a bid (morning and noon) or tid (morning, noon, and 5 p.m.) regimen. Current labelling for these formulations does not reflect this ncertainty, but as a consequence of recent review and recommendations by the Food and Drug Administrations' Cardiac and Renal Drugs Advisory Committee, revised labeling is being written.

References

1. Thadani U. Current status of nitrates in angina pectoris. *Mod Concepts Cardiovasc Dis* 1987;56:49–54.
2. Abrams JA. A reappraisal of nitrate therapy. *JAMA* 1988;256:396–401.
3. Thadani U. Medical therapy of stable angina pectoris. *Cardiol Clin* 1991;9:73–87.
4. Murrell W. Nitroglycerin as a remedy for angina pectoris. *Lancet* 1879;1:80–81.
5. Aronow WS. Use of nitrates as antianginal agents. In: Needleman P, ed. *Organic Nitrates*. New York: Springer Verlag, 1975:163–174.
6. Detry JMR, Bruce RA. Effect of nitroglycerin on "maximal" oxygen intake and exercise electrocardiogram in coronary artery disease. *Circulation* 1971;43:155–163.
7. Goldstein RE, Rosing DR, Redwood DR, et al. Clinical and circulatory effects of isosorbide dinitrate, comparison with nitroglycerin. *Circulation* 1971;43:629–640.
8. Thadani U, Parker JO. Influence of glyceryl trinitrate during supine and upright exercise in patients with angina pectoris. *Br Heart J* 1978;40:1229–1236.
9. Glancy DL, Ritcher MA, Ellis EV, et al. Effect of swallowed isosorbide dinitrate on blood pressure, heart rate and exercise capacity in patients with coronary artery disease. *Ann Intern Med* 1977;62:39–46.
10. Danahy DT, Burwell DT, Aronow WS, et al. Sustained hemodynamic and antianginal effects of high oral dose isosorbide dinitrate. *Circulation* 1977;55:381–387.
11. Markis JE, Gorlin R, Mills RM, et al. Sustained effects of orally administered isosorbide dinitrate on exercise performance of patients with angina pectoris. *Am J Cardiol* 1979;43:265–271.
12. Thadani U, Fung HL, Darke AC, et al. Oral isosorbide dinitrate in the treatment of angina pectoris: Doseresponse relationship and duration of action during acute therapy. *Circulation* 1980;62:491–502.
13. Giles TD, Iteld BJ, Quiroz T, et al. The prolonged effect of pentaerythritol tetranitrate on the exercise capacity in stable angina pectoris. *Chest* 1981;80:142–145.
14. DiBianco R, Ronan JA, Donohue DJ, et al. A new oral slow release form of isosorbide dinitrate: Effect on the hemodynamics and exercise capacity of patients with angina. *Chest* 84:707–713.

15. Thadani U, Prasad R, Hamilton SF, et al. Isosorbide-5-mononitrate in angina pectoris: Plasma concentrations and duration of effects after acute therapy. *Clin Pharmacol Ther* 1987;42:58–65.
16. Stewart DD. Tolerance to nitroglycerin. *JAMA* 1910;54:783.
17. Crandall LA, Leake CD, Loevenhart AS, et al. Acquired tolerance to and cross tolerance between the nitrates and nitric oxide esters and sodium nitrate in man. *J Pharmacol Exp Ther* 1981;41:103–119.
18. Bogaert MA, De Schaepdryver AF. Tolerance towards glyceryl trinitrate (Trinitrin) in dogs. *Arch Int Pharmacldyn Ther* 1968;171:221–224.
19. Needleman P. Tolerance to the vascular effects of glyceryl trinitrin. *J Pharmacol Exp Ther* 1970;171:98–102.
20. Schelling JL, Lasagna L. A study of cross tolerance to circulatory effects of organic nitrates. *Clin Pharmacol Ther* 1967;8:256–260.
21. Needleman P, Johnson EM Jr. Mechanism of tolerance development to organic nitrates. *J Pharmacol Exp Ther* 1973;184:709–715.
22. Thadani U, Manyari D, Parker JO, et al. Tolerance to the circulatory effects of oral isosorbide dinitrate: Rate of development and cross-tolerance to glyceryl trinitrate. *Circulation* 1980;234:166–170.
23. Yeung AC, Vekshtein VI, Krantz DS, et al. The effect of atherosclerosis on the vasomotor response of coronary arteries to mental stress. *N Engl J Med* 1991;325:1551–1556.
24. Gage JE, Hess DM, Murakami T, et al. Vasoconstriction of stenotic coronary arteries during dynamic exercise in patients with classical angina pectoris: Reversibility by nitroglycerin. *Circulation* 1986;73:865–876.
25. Palmer RMJ, Ferrige AG, Moncada S. Nitric oxide release accounts for the biologic activity of endothelial-derived relaxant factor. *Nature* 1987;327:524–526.
26. Harrison DG, Kurz MA, Quillen JE, et al. Normal and pathophysiologic consideration of endothelial regulation of vascular tone and their relevance to nitrate therapy. *Am J Cardiol* 1992;70:11B–17B.
27. Bassange E. Nitrates in different vascular beds: Nitrate tolerance, and interaction with endothelial function. *Am J Cardiol* 1992;70:23B–29B.
28. Taylor C, Hodge EM, White DA, and the Amlodipine Angina Study Group. Circadian rhythm of angina: Similarity to circadian rhythms of myocardial infarction, ischemic ST segment depression, and sudden cardiac death. *Am Heart J* 1989;118(Suppl):1098–1099.
29. Feldman RL, Pepine CJ, Conti CR. Magnitude of duration of dilation of large and small coronary arteries by nitroglycerin. *Circulation* 1981;64:324–333.
30. Ferrer MI, Bradley SE, Wheeler HO, et al. Some effects of nitroglycerin upon the splanchnie, pulmonary and systemic circulation. *Circulation* 1986;33:357–373.
31. Goldstein RE, Bennett ED, Leech GL. Effects of glyceryl trinitrate on echocardiographic ventricular dimensions during exercise in the upright position. *Br Heart J* 1979;42:245–254.
32. Bassange E, Strein K. Dose dependent effects of isosorbide-5-mononitrate on the venous, arterial and coronary arterial system of conscious dogs. *Naunyn Schmiedebergs Arch Pharmacol* 1986;334:100–104.
33. Brown BG, Bolson E, Petersen RB, et al. The mechanisms of nitroglycerine action: Stenosis dilation as a major component of drug response. *Circulation* 1981;64:1089–1097.

34. Ignarro LJ, Lippton H, Edward JC, et al. Mechanism of vascular smooth muscle relaxation by organic nitrates, nitrates, nitroprusside and nitric oxide: Evidence of the involvement of S-nitrosothiols as active intermediates. *J Pharmacol Exp Ther* 1981;218:739–749.
35. Kukovetz WR, Holzmann S. Mechanism of nitrate induced vasodilation tolerance. *Z Kardiol* 1983;72(Suppl 3):14–19.
36. Chung SJ, Fung HL. Relationship between nitroglycerin induced vascular relaxation and nitric oxide production: Problems with inhibition and tolerance development. *Biochem Pharmacol* 1983;45:157–163.
37. Murad F. Cyclic guanosine monophosphate as a mediator of vasodilation. *J. Clin Invest* 1986;78:1–5.
38. Lam JY, Chesbro JH, Fuster V. Platelets, vasoconstriction and nitroglycerin during arterial wall injury: A new antithrombotic role of an old drug. *Circulation* 1988;78:712–716.
39. Aronow WS, Chesluk MH. Evaluation of nitroglycerin in angina in patients on isosorbide dinitrate. *Circulation* 1970;42:61–63.
40. Prager, G. Buccal synchron nitroglycerin in acute treatment of angina pectoris attacks: A comparison with sublingual administered nitroglycerin in patients with stable coronary disease. *Adv Pharmacother Ther* 1982;1:137–142.
41. Parker JO, Vankoughnett KA, Farrell B. Nitroglycerin lingual spray: Clinical efficacy and dose response relation. *Am J Cardiol* 1986;57:1–5.
42. Lichstein E, Greengart A, Bolton S, et al. Efficacy of buccal synchron nitroglycerin in patients with angina pectoris. *Adv Pharmacother Ther* 1982;1:166–173.
43. Thadani U, Whitsett T, Hamilton SF. Nitrate therapy for myocardial ischemic syndrome: Current perspective including tolerance. *Curr Probl Cardiol* 1988;13:725–784.
44. Stipe AA, Fink GB. Prophylactic therapy of angina pectoris with organic nitrates: Relationship of drug efficacy and clinical experimental design. *J Clin Pharmacol* 1973;13:244–250.
45. Thadani U. Role of nitrates in angina pectoris. *Am J Cardiol* 1992;70:43B–53B.
46. Elkayam U. Tolerance to organic nitrates: Evidence, mechanisms, clinical evidence and strategies for prevention. *Ann Intern Med* 1991;114:667–677.
47. Thadani U. Nitrate tolerance: How to prevent it or minimize its effect. *Postgrad Med* 1992;91:307–318.
48. Thadani U, Fung HL, Darke AC, et al. Oral isosorbide dinitrate in angina pectoris: Comparison of duration of action and dose-response relation during acute and sustained therapy. *Am J Cardiol* 1982;49:411–419.
49. Parker JO, Vankoughnett KA, Farrell B. Comparison of buccal nitroglycerin and oral isosorbide dinitrate for nitrate tolerance in stable angina pectoris. *Am J Cardiol* 1985;56:724–728.
50. Rudolph W, Blasini R, Reinger G, et al. Tolerance development during isosorbide dinitrate treatment: Can it be circumvented? *Z Kardiol* 1983;72(Suppl 3):195–198.
51. Danahy DT, Aronow WS. Hemodynamic and antianginal effects of high dose oral isosorbide dinitrate after chronic use. *Circulation* 1977;56:205–212.
52. Schneider WU, Bussmann WD, Stahl B, et al. Dose response relation of antianginal activity of isosorbide dinitrate. *Am J Cardiol* 1984;53:700–705.
53. Thadani U, Shapiro W, DiBianco R. Effect of long term placebo therapy on angina frequency and exercise toler-ance in patient with stable angina pectoris (abstr). *Circulation* 1984;70(Suppl II):II–44.

54. Parker JO, Farrell B, Lahey KA, et al. Effect of interval between doses on the development of tolerance to isosorbide dinitrate. *N Engl J Med* 1987;316:1440–1444.
55. Bassan MM. The day long pattern of antianginal effect of long term three times daily administration of isosorbide dinitrate. *J Am Coll Cardiol* 1990;16:936–940.
56. Silber S, Vogler AC, Krause KH, et al. Induction and circumvention of nitrate tolerance applying different dosage intervals. *Am J Med* 1987;83:860–870.
57. Lee G, Mason DT, Amsterdam EA, et al. Antianginal efficacy of oral therapy with isosorbide dinitrate capsules: Prolonged benefits shown by exercise testing in patients with ischemic heart disease. *Chest* 1978;73:327–392.
58. Abshagen U, Betzien G, Endele R, et al. Pharmacokinetics of intravenous and oral isosorbide 5 mononitrate. *Eur J Clin Pharmacol* 1981;20:269–275.
59. Muller G, Uberbacher HJ, Glocke M. Double-blind, randomized, crossover study on the coronary therapeutic efficacy and tolerance of Is-5-MN (Ismo 20 mg) in comparison with IS-5-MN + metipranolol (Ismo-Disorate 20/5) and placebo. In: Cohn JN, Rittinghausen R, eds. *Mononitrates.* Berlin: Springer Verlag, 1985:221–229.
60. Rennhak U, Riebesel T, Biamino G, A double-blind cross-over study on the effectiveness and possible development of tolerance during long term therapy with isosorbide-5-mononitrate or isosorbide dinitrate slow release in coronary artery disease. In: Cohn JN, Rittinghausen R, eds. *Mononitrates*. Berlin: Springer Verlag, 1985:147–153.
61. Kohli RS, Rodrigues EA, Kardash MN, et al. Acute and sustained effects of isosorbide-5-mononitrate in stable angina pectoris. *Am J Cardiol* 1986;58:727–731.
62. Seabra-Gomes R, Alexo AM, Adalo M, et al. Comparison of the effects of a controlled-release formulation of isosorbide-5-mononitrate and conventional isosorbide dinitrate on exercise performance in man with stable angina pectoris. *Am J Cardiol* 1990;65:1308–1312.
63. Tauchert M, Jansen W, Osterspey A, et al. Dose dependence of tolerance during treatment with mononitrates. *Z Kardiol* 1983;73(Suppl 3):218–228.
64. Jones JP and the Isosorbide-5-Mononitrate Study Cooperative. Acute dose response of isosorbide-5-mononitrate in patients with angina pectoris. In: Julian DG, Rittinghausen R, Überbacher HJ, eds. Mononitrate II. Berlin: Springer Verlag, 1987:61–68.
65. Thadani U, Prasad R, Hamilton SF, et al. Usefulness of twice daily isosorbide 5 mononitrate in preventing development of tolerance in angina pectoris. *Am J Cardiol* 1987;60:477–482.
66. Thadani U, Friedman R, Jones JP, et al. Nitrate tolerance eccentric versus concentric twice daily therapy with isosorbide-5-mononitrate in angina pectoris (abstr). *Circulation* 1989;2:216.
67. Thadani U and IS-5-MN Study Group. Isosorbide-5-mononitrate [IS-5-MN] in angina pectoris: Efficacy of am and pm doses, lack of tolerance and zero hour effect during eccentric bid therapy (abstr). *Circulation* 1991;84(Suppl II):II–730.
68. Thadani U, Bittar N. Effects of 8:00 am and 2:00 pm doses of isosorbide-5-mononitrate during twice daily therapy in stable angina pectoris. *Am J Cardiol* 1992;70:286–292.
69. Thadani U, Devane PJ. Efficacy of isosorbide mononitrate in angina pectoris. *Am J Cardiol* 1992;70:67G–71G.
70. Thadani U, Maranda C, Amsterdam E, et al. Lack of pharmacologic tolerance and rebound angina pectoris during

twice daily therapy with isosorbide-5-mononitrate. *Ann Intern Med* 1994;120:353–359.
71. Thadani U, Whitsett T. Relationship of pharmacokinetics and pharmacodynamic properties of organic nitrates. *Clin Pharmacokinet* 1988;15:32–43.
72. Parker JO and the Isosorbide-5-Mononitrate Study Group. Eccentric dosing with isosorbide-5-mononitrate in angina pectoris. *Am J Cardiol* 1993;72:871–876.
73. Demots H, Glasser SP. Intermittent transdermal nitroglycerin therapy in the treatment of chronic stable angina. *J Am Coll Cardiol* 1989;13:786–795.
74. Thadani U, Hamilton SF, Olson E, et al. Duration of effects and tolerance of slow release isosorbide-5-mononitrate for angina pectoris. *Am J Cardiol* 1987;59:756–762.
75. Nyberg G, Carlene P, Lindstrom E, et al. The effect of isosorbide-5-mononitrate (IS-5-MN) durules on exercise tolerance in patients with exertional angina pectoris. A placebo controlled study. *Eur Heart J* 1986;7:836–842.
76. Wisenberg G, Rucks C, Nichol P, et al. Sustained effect of and lack of development of tolerance to controlled-release isosorbide-5-mononitrate in chronic stable angina pectoris. *Am J Cardiol* 1989;64:569–576.
77. Beyerle A, Reinger G, Rudolph W. Long-acting marked antiischemic effect maintained during long term interval treatment with once daily isosorbide-5-mononitrate in sustained release form. *Am J Cardiol* 1990;65:1434–1437.
78. Parker JO. Controlled release isosorbide-5-mononitrate in angina pectoris: A comparison with standard formulation isosorbide dinitrate. *Can J Cardiol* 1991;7:125–130.
79. Chrysant SG, Glasser SP, Bittar N, et al. Efficacy and safety of extended release isosorbide-mononitrate for stable effort angina pectoris. *Am J Cardiol* 1993;72:1249–1256.
80. Steering Committee, Transdermal Nitroglycerin Cooperative Study. Acute and chronic antianginal efficacy of continuous 24-hour application of transdermal nitroglycerin. *Am J Cardiol* 1991;63:1263–1273.
81. Zimrin D, Reichek N, Bogin KT, et al. Antianginal effects of intravenous nitroglycerin over 24 hours. *Circulation* 1988; 77:1376–1384.
82. Berkenboom GM, Sobolski JC, Degre SA. Oral sustained-release nitroglycerin in chronic stable angina. A multi-center, double blind, randomized crossover trial. *Am J Cardiol* 1984;53:15–17.
83. Degre SG, Strappart GM, Sobolski JC, et al. Effect of oral SRNG on exercise capacity in angina pectoris. *Am J Cardiol* 1983;51:1595–1598.
84. Winsor T, Berger H. Oral nitrolgycerin as a prophylactic antianginal drug: Clinical, physiologic and statistical evidence of efficacy based in a three phase experimental design. *Am Heart J* 1975;90:611–626.
85. Ghio S, Deservi S, Perotti S, et al. Differential susceptibility to the development of nitroglycerin tolerance in the arterial and venous circulation in humans: Effects of N-acetylcysteine administration. *Circulation* 1992;86:798–802.
86. Packer M, Lee WH, Kessler PD, et al. Prevention and reversal of nitrate tolerance in patients with congestive heart failure. *N Engl J Med* 1987;317:799–804.
87. Parker JO, Farrell LB, Lahey KA, et al. Nitrate tolerance: The lack of effect of N-acetylcysteine. *Circulation* 1987;76:572–576.
88. Dupuis J, Lalonde G, Lemieux R, et al. Tolerance to intravenous nitroglycerin in patients with congestive heart failure: Role of increased intravascular volume, neurohumoral activa-

tion and lack of prevention with N-acetylcysteine. *J Am Coll Cardiol* 1990;16:923–931.
89. Lis Y, Bennett D, Lambert G, et al. A preliminary double blind study of intravenous nitroglycerin in acute myocardial infarction. *Intensive Care Med* 1984;10:179–184.
90. Boesgaard S, Poulsen E, Aldershvile J, et al.Nitrate tolerance is not associated with depletion of arterial or venous sulfhydryl levels in vivo (abstr). *Circulation* 1992;86(Suppl I):488.
91. Katz R, Levy WS, Buff L, et al. Prevention of nitrate tolerance with angiotensin converting enzyme inhibitors. *Circulation* 1991;83:1271–1276.
92. Dakak N, Makhoul N, Flugelman MY, et al. Failure of captopril to prevent nitrate tolerance in congestive heart failure secondary to coronary artery disease. *Am J Cardiol* 1990;66:608–612.
93. Sussex BA, Campbell NR, Rain MK. Nitrate tolerance is modifid by diuretic treatment. *Circulation* 1990;80(Suppl III):200.
94. Parker JO, Farrell B, Fenton T, et al. Effects of diuretic therapy on the development of tolerance during continuous therapy with nitroglycerin. *J Am Coll Cardiol* 1992;20:616–620.
95. May DC, Pompa JJ, Black WM, et al. In vivo induction and reversal of nitroglycerin tolerance in human coronary arteries. *N Engl J Med* 1987;317:805–809.
96. Thadani U. Factors influencing tolerance in nitrate therapy: A review. In: Julian DG, Rittinghausen R, Überbacher HJ, eds. *Mononitrate II*. Berlin: Springer-Verlag, 1987:82–98.
97. Thadani U. Nitrate tolerance at the cellular level: Is it real? Proceedings of a Symposium. Heart Disease and the Endothelium: Impact of Treatment. *Perspect Cardiol* 1992;Suppl:23–27.
98. Thadani U. Combination therapy for the angina patient. *Am Coll Cardiol* 1989;5:15–19.
99. Akhras S, Jackson G. Efficacy of nifedipine and isosorbide mononitrate in combination with atenolol in stable angina. *Lancet* 1991;338:1036–1039.

Ointments and Transdermal Nitroglycerin Patches for Stable Angina Pectoris

Udho Thadani and Raymond J. Lipicky

Summary. Nitroglycerin (NTG) ointment is used for the prophylaxis against angina pectoris, but there are no data to support its effectiveness during long-term therapy. Continuous, once-daily application of isosorbide dinitrate cream produces tolerance with complete loss of efficacy within 1 week. Nitroglycerin patches are very popular and continuous once-daily application is still claimed by some investigators to provide 24 hour antiischemic and antianginal efficacy. This claim is based on data from postmarketing studies in a very large number of patients and placebo-controlled studies in smaller groups of patients from Italy, Yugoslavia, Greece, and Germany. In contrast, studies from the United States, Canada, England, and some centers in Germany have failed to show superiority of patches over placebo during continuous therapy. This controversy was addressed by the NTG cooperative study group, in which a total of 562 patients who were responders to sublingual nitroglycerin were studied. Patients received either placebo or NTG patches delivering low (15–30 mg/24 hr), moderate (45–60 mg/24 hr), or large (75 and 105 mg/24 hr) amounts of NTG. Four hours after the initial application, NTG patches increased exercise duration compared to placebo, but this beneficial effect had disappeared by 24 hours. Furthermore, after 8 weeks of continuous therapy, none of the NTG patches were superior to placebo, whether patients were or were not taking concomitant beta-blockers. Therefore, current opinion is that continous therapy with NTG patches produces pharmacologic tolerance and is ineffective. Pharmacologic tolerance can be minimized when patches are applied every morning and removed after 10–12 hours at night. However, patches delivering >15 mg NTG/24 hr are required to maintain an increased exercise duration for up to hour 8 after the patch application. Intermittent therapy with patches, however, may lead to rebound nocturnal angina in some patients. Also, intermittent therapy with patches has been associated with worsening of exercise performance in the morning prior to the patch renewal, compared to therapy with

The opinions expressed here are those of the authors and should not be taken as those of FDA.

placebo patches. This has been referred to as the zero-hour effect and probably represents a rebound phenomenon following nitrate withdrawal. Patients experiencing either nocturnal or early morning angina during intermittent therapy with patches should either be switched to oral long-acting nitrates or should in addition be treated with a beta-blocker, provided there are no contraindications to beta-blocker treatment.

Key Words. **nitroglycerin ointment, nitroglycerin patches, isosorbide dinitrate cream, angina pectoris, tolerance, continuous versus intermittent treatment**

Nitrates have been used by the transcutaneous route in the form of nitroglycerin (NTG) ointment for almost 50 years to treat patients with ischemic heart disease [1,2]. Transcutaneous administration of nitrates avoids first-pass hepatic metabolism in contrast to oral formulations of NTG and isosorbide dinitrate, which undergo extensive first-pass hepatic clearance [3,4]. However, NTG ointment has to be applied to a large surface area, has variable intraindividual and interindividual absorption [3], and the duration of effect after the first application is only 3–8 hours [5–11], necessitating three- or four-times-a-day application if NTG ointment was to be used for chronic administration.

In order to increase the systemic bioavailability and provide constant delivery of NTG for 24 hours or longer, transdermal delivery systems (patches and disks) were formulated and were shown to provide fairly constant NTG plasma concentrations throughout the 24 hours [12,13]. Repeated once-daily application was conditionally approved in 1983 in the United States for 24 hour prophylaxis of angina pectoris by the Food and Drug Administration on the basis of bioavailability data. Early reports [14,15] of beneficial effects of patches were generally thought to have been confirmed in open-label postmarketing studies [16,17], and patches gained wide acceptance both with physicians and patients alike. Subsequent placebo-controlled studies, however, suggested that NTG patches did not provide 24-hour antianginal efficacy [18–30] and that tolerance reported previously with oral nitrates [31] developed rapidly. Other reports, however, especially from Italy, Yugoslavia, and Greece showed that continuous therapy with patches was effective for 24 hours or longer [32–51]. Thus, controversy regarding the usefulness of patches prevailed for several years and to resolve this controversy, a dose-response study with the patches was conducted in the United States under the auspices of the Food and Drug Administration (FDA). The results of that study published in 1991 showed a lack of beneficial effect during long-term therapy with active patches delivering 15–105 mg NTG/24 hr compared to placebo patches [52]. However, despite the negative multicenter cooperative study, a meta-analysis published in 1993 concluded that continuous therapy with NTG patches was effective during long-term therapy [53]. In the United States a change in the labelling in 1991 recommended intermittent rather than continuous therapy with patches. This review and recommendations for appropriate therapy with ointment and patches is based on the results of published placebo-controlled studies.

Nitroglycerin Ointment

Nitroglycerin ointment is still widely used to treat patients with stable angina pectoris. Upon the first application, ointment exerts significant hemodynamic, antiischemic, and antianginal effects [5–11]. There is, however, little information regarding the long-term efficacy of the ointment. In open-label studies, NTG ointment reduced the frequency of anginal attacks and increased exercise duration [1,6–9]. However, similar beneficial effects may be obtained with placebo therapy in patients with angina pectoris [25,52].

In a double-blind study, the first application of NTG ointment improved exercise duration for up to 6–8 hours in comparison to placebo [5]. On the basis of this report, four-times-a-day application of NTG ointment during long-term therapy was recommended. In a recent report, attenuation of effects on improvements in exercise duration was observed after 3 weeks of twice-daily application of 2% NTG ointment to a surface area of 2.5 cm^2 (low dose) and 6.5 cm^2 (high dose) [54]. There are no other controlled studies reported. Thus published data are inadequate and definite recommendations as to how much and how often the ointment should be used during long-term therapy cannot be made at present.

Isosorbide-Dinitrate Cream

Isosorbide-dinitrate ointment is commercially available, but its continuous application is associated with the development of tolerance, with complete loss of antianginal effects within a week [55,56]. Whether intermittent use of ointment will be useful in treating patients with angina is not known; therefore, this preparation cannot be recommended to treat patients with stable angina pectoris.

Nitroglycerin Patches

Continuous daily application

The enormous success of NTG patches as an antianginal agent is based on data showing a marked reduction in daily anginal attacks during daily life in open-label studies [15]. Postmarketing surveys [16,17] involving several thousand patients showed a 80–90% reduction in angina frequency during daily activities. Similar observations were made in open-label studies where effects of therapy on exercise performance were compared to baseline.

Effects on exercise duration

Earlier studies. After the application of the first patch, delivering 5–45 mg of NTG per 24 hours, exercise duration to the onset of angina and total exercise duration improved at hours 2 and 4 after the application of the patches [18–30,57–60]. There are reports, especially from Italy, showing an increase in exercise duration at hours 16–18 and at hour 24 after the application of the patches delivering 5–20 mg per 24 hours [32–51]. In contrast, several well-designed, placebo-controlled studies performed in the United States, Canada, Australia, and England have failed to show an improvement in exercise duration or reduction in myocardial ischemia during exercise either using patches delivering 5–10 mg of (low dose) or 45 mg/24 hr of NTG (medium dose) [18–30,54–60].

Recent studies. Three recent studies from Europe, one using a crossover design and another a parallel design, showed that continuous therapy with NTG patches does not produce tolerance [48,49,51]. In these studies, exercise tolerance was shown to increase at 2, 4, and 24 hours after first application, as well as after patch renewal.

In contrast, in a large-patient population-based multicenter parallel-designed placebo-controlled study from the United States, in which 562 patients were studied, patch sizes delivering NTG 15 mg/24 hr increased exercise duration at hour 4, but was not superior to placebo at hour 24 (Figure 1) [52]. More importantly, after 8 weeks of continuous therapy with patches once a day, there was no benefit seen in any of the active patch groups compared to placebo (Figure 2) despite increase of doses up to 105 mg/24 hr. A

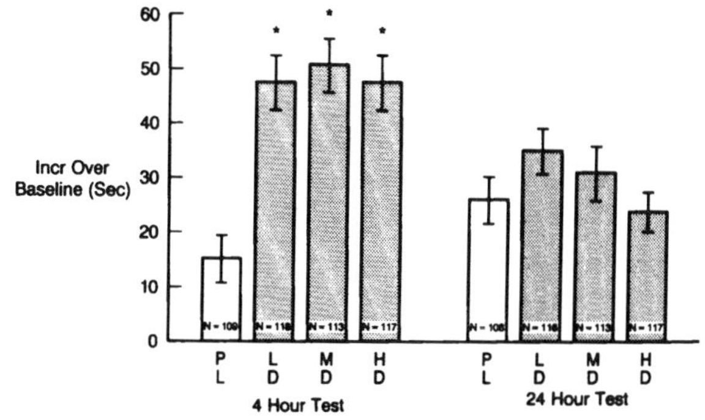

Fig. 1. *Usefulness of initial application of nitroglycerin patches in angina pectoris. Compared to double-blind placebo, treadmill exercise duration increased significantly at 4 hours, but this effect was no longer evident 24 hours after NTG patch application. The exercise tolerance times shown in the figure were obtained from patients who were randomly assigned to low-dose (<15–30 mg/24 hr), medium-dose (45 and 60 mg/24 hr), or high-dose (75 and 105 mg/24 hr) groups. However, the first patch applied (shown in the figure) was 15 mg/24 hr. *p < 0.05 over placebo. (From Steering Committee [52], with permission.)*

marked improvement in exercise tolerance during long-term placebo therapy was seen in this study (Figure 2).

In a recent review from Europe [53], it was suggested that the patients in the above NTG cooperative study were unstable at baseline, as they showed an increase in exercise duration during double-blind placebo therapy. In fact, all patients in that study had reproducible exercise-limiting angina at baseline, and all patients entered were required to be sublingual NTG responders (by exercise tolerance test during

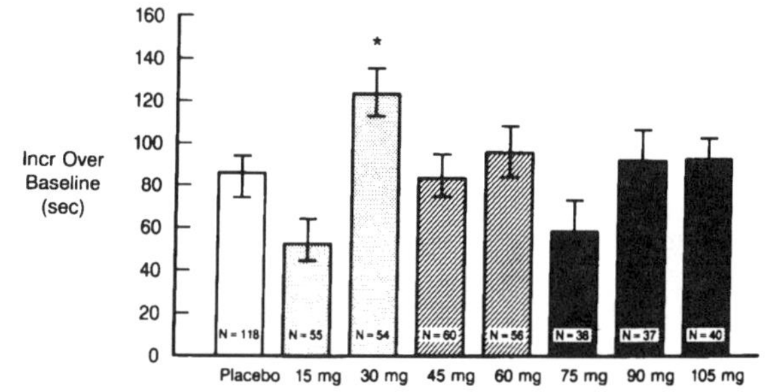

Fig. 2. *Usefulness of continuous daily application of nitroglycerin patches in angina pectoris. After 8 weeks of double-blind therapy, all groups including placebo showed an increase in exercise duration over baseline. Only the 30 mg/24 hr group demonstrated a nominally significant increase that was greater than the placebo response. *p < 0.05 over placebo. (From Steering Committee [52], with permission.)*

qualifications). The improvement in exercise tolerance over time in the placebo group is not unique to that study and has been reported previously in similar parallel-design studies. It has also been suggested [53] that in the NTG cooperative study [52], exercise performance did not improve as many of the patients took concomitant beta-blocker therapy. In reality, less than 50% of the patients were on beta-blocker therapy. Analysis of the data showed that tolerance developed whether patients were or were not taking beta-blockers [52]. Increased uptake of nitroglycerin from the skin during exercise [71] as an explanation for variable results of different studies [53] is improbable in view of the results of coop study [52].

Thus, from the available published literature, it would appear that controversy is still present and that one could, on the basis of data and with respect to the questions of tolerance, support more than one point of view. We suggest that there is no need for debate nor for continued controversy.

That the first application of a 15 mg/24 hr (or greater) NTG patch, after a nitrate-free interval of at least 8 hours, is capable of producing a greater increase in exercise tolerance than is placebo is without doubt. The transdermal nitroglycerin cooperative study [52], the study that also showed tolerance to doses as great or 105 mg/24 hr, clearly found an effect of the first patch (15 mg/24 hr) applied, as have many others.

That tolerance occurs is equally amply demonstrated. Tolerance to the effects of NTG is regularly and reproducibly seen in a variety of in vitro test system [61–65]. Tolerance is the basis of the well-reported NTG workers' syndrome [66,67]. Moreover, in symptom-limited exercise studies conducted using chronic administration of NTG, isosorbide dinitrate, and isosorbide mononitrate, tolerance has been clearly demonstrated [18–25,31,52,68–70]. The largest single trial ever conducted with NTG patches [52] clearly showed that the increase, compared to placebo, in exercise tolerance that was observed upon application of the first patch could not be observed again, despite increasing the dose almost by one order of magnitude. Thus, tolerance clearly has been demonstrated, and it seems futile to suggest that tolerance does not occur. It does.

Why, then, can there be some apparently adequate and well-controlled, symptom-limited exercise tolerance trials that find continuous therapy with NTG patches to be superior to placebo? "There is no clear-cut answer." One possibility is that so many studies have been done that there are occasional apparent favorable reports that are merely outliers. One of these occasions is the transdermal nitroglycerin cooperative study [52]. In that study, after 8 weeks the 30 mg/24 hr group had a "statistically significant" ($p < 0.05$) increase in exercise duration compared to placebo. But the overall study overwhelmingly showed that tolerance to nitroglycerin was total; no effects were seen for any other dose. The finding of superiority over placebo for the 30 mg/24 hr group was almost certainly not real and resulted from multiple comparisons made, that is, one of a great many tests was positive, and therefore statistical tests need adjustment for multiple comparisons.

Similarly, it should not be surprising for occasional studies to show a statistically significant difference compared to placebo, even when there is, in fact, no difference. Although in studies that have found tolerance the development of tolerance seems to occur within 24 hours, this time course is well defined for only the 15 mg/24 hr NTG patch. Studies with other doses of nitroglycerin patches and with other nitrate preparations, however, have shown that longer time courses of 1–4 weeks lead to development of significant tolerance (data on file at the U.S. Food and Drug Administration). One should therefore not accept studies of 1 week and shorter as a sufficient duration to have evaluated tolerance.

From the published data, including the most convincing data of the large cooperative study [52], the following conclusions can be made regarding continuous therapy with patches:

1. Continuous therapy with NTG patches led to the development of rapid tolerance within 24 hours, if not earlier.
2. Tolerance could not be avoided or overcome by either keeping the dose the same or by increasing the dose to as high as 105 mg/24 hr.
3. Tolerance developed whether patients were or were not taking concomitant beta-blocker therapy.

Thus, continuous therapy with patches cannot be recommended.

Effects on anginal episodes

Arguments have been made that antianginal drugs are prescribed primarily to abolish or reduce the number of anginal episodes that patients experience during daily activities and that the results of exercise tests have little clinical relevance. Published studies have reported beneficial effects of continuous therapy with patches on the daily anginal attack rate [16,17]. In the large NTG cooperative study, a reduction in anginal attacks was not observed in any of the groups receiving NTG patches compared to placebo [52]. A subgroup post hoc analysis did show a reduction in frequency of angina attacks in patients who experienced more than seven attacks of angina per week during single-blind placebo therapy [52]. However, caution is required in drawing any conclusions from this post hoc analysis, which was only one of many; other equally plausible analyses of the data found no statistically significant differences between patch and placebo. In another large study [25], there was no difference in the frequency of anginal attacks or sublingual NTG consumption during long-term continuous therapy with patches delivering 5 mg NTG/24 hr compared to placebo patches.

Patch size and efficacy

It has been suggested that patches delivering larger quantities of NTG/24 hr should be effective for up to 24 hours. The NTG cooperative study, however, which used patches delivering up to 105 mg of NTG/24 hr, did not show effectiveness at any time during continuous long-term therapy [52].

Cross-tolerance to sublingual nitroglycerin

In an earlier study with patches delivering 15 mg of NTG/24 hr, no cross-tolerance to sublingual NTG was reported [19]. In a recent study, after 24 hours of therapy with intravenous NTG that provided constant plasma concentrations of 5 ng/ml, there was less increase in exercise duration after sublingual NTG than was seen in a placebo group not receiving intravenous NTG, suggesting development of cross-tolerance to sublingual NTG [72]. A similar observation was made in patients who received patches delivering high doses of NTG per 24 hours in the NTG cooperative study [52].

Intermittent Therapy with Nitroglycerin Patches

With the recognition that tolerance developed rapidly during continuous application of patches, and the knowledge from previous animal and human studies that tolerance would disappear after a nitrate-free interval, the usefulness of intermittent therapy with patches has been evaluated [73–85]. In contrast to the continuous patch studies, outcomes were quite uniform (except for doses less than 15 mg/24 hr); all studies found intermittent therapy to be effective. In general, the patches were applied for 12 hours during every day and removed at night. An improvement in exercise duration $3\frac{1}{2}$–4 hours after patch application

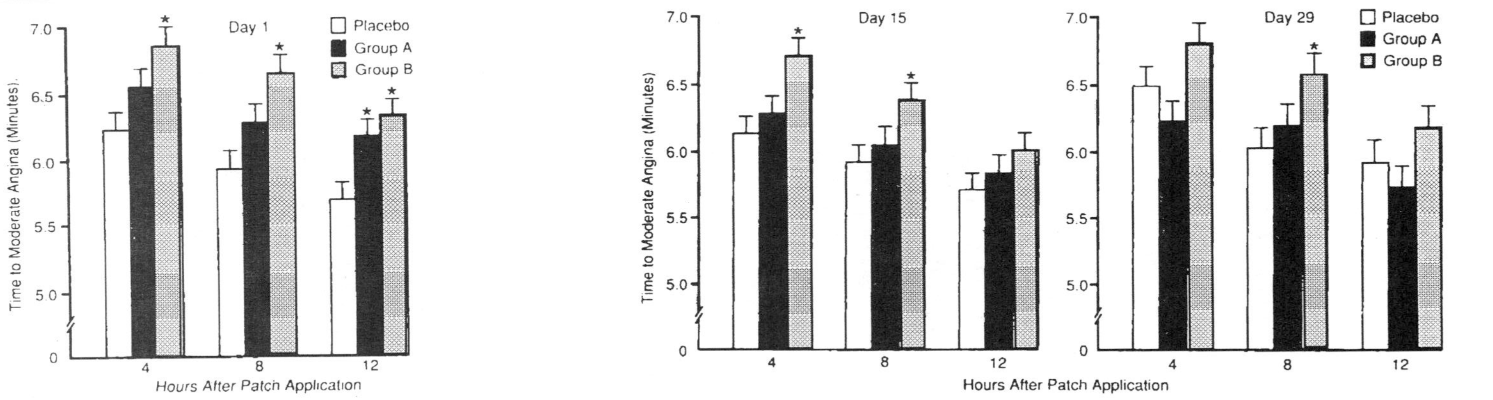

Fig. 3. *Usefulness of intermittent (12 hours on and 12 hours off) treatment with nitroglycerin patches. After 29 days of treatment, only the medium NTG dose group B (15–20 mg/24 hr) showed an increase in exercise duration over placebo at 8 hours after the application of the patch. No greater effects over placebo were seen with the low NTG dose, group A (5–10 mg/24 hr). *$p < 0.05$. (From DeMots et al. [82], with permission.)*

during intermittent therapy was reported in several studies [73–81,83] after 7 days of therapy. In a placebo-controlled, parallel study [82], patches delivering less than 15 mg of NTG/24 hr produced an initial increase in exercise duration after the first application, but the beneficial effect was lost during long-term therapy (Figure 3). Patches delivering 15–30 mg of NTG per 24 hours increased exercise duration after the first patch, and at 4 and 8 hours after long-term therapy, but the improvement in exercise performance beyond 8 hours was not clinically meaningful [82].

Patch-off therapy

Whether removal of patches for less than 12 hours would prevent development of tolerance was addressed in several studies. Removal of patches for only 4 hours was insufficient to prevent tolerance [76], and even when patches were worn for only 16 hours every day (8 hours off), tolerance developed within 12 hours of patch application [78]. Possible problems not well enough described to date include (a) Rebound increase in nocturnal angina after patch removal and (b) a zero-hour negative effect on exercise performance.

Rebound increase in nocturnal angina during intermittent patch therapy

During intermittent therapy with patches, a rebound increase in nocturnal angina during the patch-free interval was reported in 8% of patients in the large, cooperative parallel-group study [82]. Some smaller studies [83,84] have failed to show a rebound increase in nocturnal angina during the patch-free interval. Further data are needed to address this issue.

Zero-hour effect

In a recent study [82], after 15 and 29 days of intermittent NTG patch therapy, exercise duration in the morning prior to the renewed application of the patches was reduced compared to baseline values and placebo-treated patients (Figure 4). This reduction in exercise tolerance, 12 hours following patch removal, and before the renewal of the patch (zero-hour effect), may be due to a reflex increase in vasomotor tone in the early hours of the morning. The patient is, therefore, not protected at this time and may be at a disadvantage at a time when there is a higher incidence of stroke, myocardial infarction, and sudden death.

The following conclusions can be derived from the published data on intermittent therapy with patches:

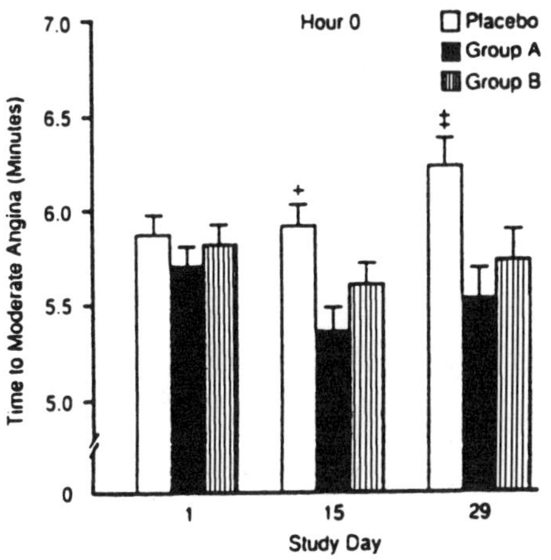

Fig. 4. *Adverse effect of intermittent (12 hours on and 12 hours off) treatment with nitroglycerin patches on exercise duration before patch application in angina pectoris. Total exercise time increased in placebo group but not in low-dose group A (5–10 mg/24 hr) or medium-dose group B (15–20 mg/24 hr) NTG over baseline. Compared to double-blind placebo, exercise duration was significantly lower in the low- and medium-dose NTG groups after 15 and 29 days of therapy. *$p < 0.05$, ‡$p < 0.01$ compared to placebo. (From DeMots et al. [82], with permission.)*

1. Patches delivering less than 15 mg of NTG/24 hr are not effective during long-term therapy.
2. Patches delivering 15–30 mg NTG/24 hr increase exercise duration for up to 8 hours during long-term therapy.
3. Partial tolerance develops as early as 12 hours, after the application of the patches.
4. There may be a rebound increase in nocturnal angina during nitrate-free interval remains a concern.
5. Intermittent therapy with patches may be associated with worsening of exercise performance in the morning prior to the renewal of the active patch (zero-hour effect).
6. It may be prudent to warn patients against engaging in strenuous activities early in the morning and for 1–2 hours after the patch application every morning.

Combination therapy of nitroglycerin patches with beta-blockers and/or calcium channel blockers

Intermittent therapy with patches does not provide 24 hour antianginal prophylaxis, may expose the patient to nocturnal angina, and may be associated with a reduction in exercise duration prior to patch renewal. Concomitant treatment with a beta-blocker that has been shown to increase exercise duration and to reduce exercise-induced myocardial ischemia also attenuates the surge of early morning ischemia [86] and is recommended by some investigators [87,88].

An alternative to concomitant beta-blocker therapy is the concomitant use of a calcium channel blocker. There are, however, no controlled studies published regarding the safety and efficacy of this combination. Since both nitrates and dihydropyridine calcium blockers (such as nifedipine) are potent vasodilators, the potential for a marked fall in blood pressure and an increase in heart rate during such a combination therapy remains a possibility. On theoretical grounds, drugs such as diltiazem or verapamil, which do not increase heart rate and provide less vasodilation compared to dihydropyridine agents, might be more suitable for combination therapy with nitrates [87,88].

Conclusions

Continuous therapy with patches cannot be recommended, as it produces rapid tolerance with complete loss of efficacy after a few (perhaps as little as 2) days. Intermittent therapy with patches, applied every morning and removed at night, does exert persistent antianginal effects, but patch sizes delivering >15 mg NTG/24 hr are recommended. Patients should be warned against a possible increase in nocturnal angina attacks during the nitrate-free interval. Intermittent therapy with patches also may lead to a deterioration in exercise duration prior to the renewal of the patch in the morning, and patients should be warned against engaging in strenuous activities first thing in the morning. Patients experiencing either nocturnal or early morning angina during intermittent therapy with patches should either be switched to oral long-acting nitrates or should be treated in addition with a beta-blocker, provided there are no contraindications to beta-blocker treatment.

References

1. Davis JA, Wiesel BH. The treatment of angina pectoris with a nitroglycerin ointment. *Am J Med Sci* 1955;230:259–263.
2. Demma FJ, Wilson RR. The transdermal administration of nitrates: An overview. *Angiology* 1983;34:1–10.
3. Bogaert MG. Clinical pharmacokinetics of glyceryl trinitrate following the use of systemic and topical preparations. *Clin Pharmacokin* 1987;12:1–11.
4. Needleman P, Lang S, Johnson EM Jr. Organic nitrates: Relationship between biotransformation and rational

angina pectoris therapy. *J Pharmacol Exp Ther* 1972; 181:485–497.

5. Reichek N, Goldstein RE, Redwood DR, et al. Sustained effects of nitroglycerin ointment in patients with angina pectoris. *Circulation* 1974;50:348–352.
6. Davidov ME, Mrcozek WJ. The effect of nitroglycerin ointment on exercise capacity in patients with angina pectoris. *Angiology* 1976;27:205–211.
7. Karsh DL, Umbach RE, Cohen LS, et al. Prolonged benefit of nitroglycerin ointment on exercise tolerance in patients with angina pectoris. *Am Heart J* 1978;96:588–595.
8. Awan NA, Miller RR, Maxswell KS, et al. Cardiocirculatory and antianginal actions of nitroglycerin ointment. *Chest* 1978;73:14–18.
9. Davidov ME. Cutaneous administration of nitroglycerin in patients with angina pectoris. *Angiology* 1981;32:16–20.
10. Hubner PJB, Jones PRM, Galer IAR. Assessment of dermal glyceryl trinitrate and isosorbide dinitrate for patients with angina pectoris. *Br Med J* 1985;290:514–516.
11. Armstrong PW, Armstrong JA, Marks GS. Pharmacokinetics hemodynamic studies of nitroglycerin ointment in congestive heart failure. *Am J Cardiol* 1980;48:670–675.
12. Müller P, Imhof PR, Burkart F, et al. Human pharmacological studies of a new transdermal system containing nitroglycerin. *Eur J Clin Pharmacol* 1982;22:473–480.
13. Chien YW. Pharmaceutical considerations of transdermal nitroglycerin delivery: The various approaches. *Am Heart J* 1984;108:207–216.
14. Thompson RH. The clinical use of transdermal delivery devices with nitroglycerin. *J Vasc Dis*, 1983:23–31.
15. Georgopoulos AJ, Markis A, Georgiadis H. Therapeutic efficacy of a new trandermal system containing nitroglycerin in patients with angina pectoris. *Eur J Clin Pharmacol* 1982;22:481–485.
16. Letzel H, Johnson LC. Advances in nitrate therapy for ischemic heart disease: Results of a multicenter study with nitroderm TTS. *Selecta* 1983;16(Suppl):13–14.
17. Bridgman KM, Carr M, Tattersall AB. Post-marketing surveillance of the transiderm-nitro patch in general practice. *J Int Med Res* 1984;12:40–45.
18. Crean PA, Ribeiro P, Crea F, et al. Failure of trandermal nitroglycerin to improve chronic stable angina: A randomized placebo-controlled, double-blind, double crossover trial. *Am Heart J* 1984;108:1494–1500.
19. Parker JO, Fung HL. Transdermal nitroglycerin in angina pectoris. *Am J Cardiol* 1984;54:471–476.
20. Reichek N, Priset C, Zimrin D, et al. Antianginal effects of nitroglycerin patches. *Am J Cardiol* 1984;54:1–7.
21. Sullivan M, Savvides M, Abouantoun S, et al. Failure of transdermal nitroglycerin to improve exercise capacity in patients with angina pectoris. *J Am Coll Cardiol* 1985;5:1220–1223.
22. James MA, Walker PR, Papouchado M, et al. Efficacy of transdermal glyceryl trinitrate in the treatment of chronic stable angina pectoris. *Br Heart J* 1985;53:631–635.
23. Reiniger G, Rudolph W. Therapie der koronaren herzerkrankung mit nitroglycerinflastern. *Herz* 1985;10:305–311.
24. Thadani U, Hamilton SF, Olson E, et al. Transdermal nitroglycerin patches in angina pectoris. *Ann Intern Med* 1986;105:485–492.
25. Fletcher A, McLoone P, Bulpitt C. Quality of life on angina therapy: A randomized controlled trial of transdermal glyceryl trinitrate against placebo. *Lancet* 1988;ii:4–8.

26. Metelitsa VI, Martsevich SY, Piotrovskii VK. New transdermal and transmucosal nitroglycerin delivery systems in patients with ischemic heart disease. *Eur J Clin Pharmacol* 1987; 32:5–10.
27. Karlson BW, Henning R. Comparison of nitroglycerin-TTS and long-acting nitroglycerin tablets in the treatment of angina pectoris: A double-blind controlled study. *Clin Cardiol* 1987;10:573–577.
28. Frishman WF, Giles T, Greenberg S, et al. Sustained high-dose nitroglycerin transcutaneous patch therapy in angina pectoris: Evidence for attenuation of effect over time. *J Clin Pharmacol* 1989;29:1097–1105.
29. Bassan MM, Rogel S. A comparison of the day-long antianginal effectiveness of nitroglycerin patches with that of three-times-daily isosorbide dinitrate: A double-blind study using dose titration. *Eur Heart J* 1992;13:1265–1270.
30. Reifart N, Bussman WD. Placebo-kontrollierte doppelblind-studie zur antiischëmischen wirksamkeit und wirkdauer von transdermalem depotnitrat. *Z Kardiol* 1986;75(Suppl 3):83–85.
31. Thadani U, Fung HL, Darke AC, et al. Oral isosorbide dinitrate in angina pectoris: Comparison of duration of action and dose response relation during acute and sustained therapy. *Am J Cardiol* 1982;49:411–419.
32. Cerri B, Grasso F, Cefis M, et al. Comparative evaluation of the effect of two doses of nitroderm TTS on exercise-related parameters in patients with angina pectoris. *Eur Heart J* 1984;5:710–715.
33. Naafs MA, De Boer AC, Koster RW, et al. Exercise capacity with transdermal nitroglycerin in patients with stable angina pectoris. *Eur Heart J* 1984;5:705–709.
34. Schneider W, Michel O, Kaltenbach M, et al. The antianginal effect of transdermally-applied nitroglycerin as a function of the patch size. *Dtsch Med Wochenschr* 1985;110:87–91.
35. Sellier P, Audouin P, Payen B, et al. Therapeutic efficacy of transcutaneously absorbed nitroglycerin evaluated by exercise testing in angina pectoris. *Cardiovasc Rev Rep* 1985;6:80–88.
36. Hollenberg M, Go M. Clinical studies with transdermal nitroglycerin. *Am Heart J* 1984;108:223–231.
37. Martines C. Comparison of the prophylactic antianginal effect of two doses of nitroderm TTS in out-patients with stable angina pectoris. Comparison of the prophylactic antianginal effect of two doses of nitroderm TTS in outpatient with stable angina pectoris. *Curr Ther Res* 1984;36:483–489.
38. Rezakovlć DE, Pavičić L, Majačić M. A randomized placebo controlled, double-blind crossover trial of transdermal nitroglycerin in stable angina pectoris. *Eur Heart J* 1988;9(Suppl A):73–81.
39. Imhof PR, Müller P, Garnier B. Nitroderm TTS versus oral isosorbide dinitrate: A double-blind trial in patients with angina pectoris. *Acta Therapeutica* 1985;11:155–168.
40. Scardi S, Pivotti F, Fonda F, et al. Effect of a new transdermal therapeutic system containing nitroglycerin on exercise capacity in patients with angina pectoris. *Am Heart J* 1985;110:547–551.
41. Thompson R. Influence of transdermal nitrates on exercise capacity in patients with stable angina. *J Vasc Dis* 1986;38:448–453.
42. Pucci P, Zambaldi G, Cerisano G, et al. Evaluation of a new preparation of transdermal nitroglycerin for patients with angina of effort. *G Ital Cardiol* 1983;13:167–171.
43. Shell WE, Dobson D. Dissociation of exercise tolerance and

total myocardial ischemic burden in chronic stable angina pectoris. *Am J Cardiol* 1990;66:42–48.

44. Muiesan G, Agabiti-Rosei E, Muisan L. A multicenter trial of transdermal nitroglycerin in exercise-induced angina: Individual antianginal response after repeated administration. *Am Heart J* 1986;112:233–238.
45. Gibelli G, Negrini M, Bruno AM, et al. Chronic effects of transdermal nitroglycerin in stable angina pectoris: A within patient, placebo controlled study. *Int J Clin Pharmacol Ther Toxicol* 1989;27:436–441.
46. Greco R, DéAlterio D, Schiattarella M. Efficacy of a new transdermal nitroglycerin patch (Deponit 10) for stable angina pectoris. *Am J Cardiol* 1988;61:44E–51E.
47. Caru B, Pirelli S, Ferrantini M, et al. The effects on exercise tolerance of a new transdermal therapeutic system containing nitroglycerin, in patients with stable angina pectoris. *Eur Heart J* 1988;9(Suppl A):105–111.
48. Rezaković DE, Pavičić L, Majačić M. A randomized placebo controlled, double-blind, crossover trial of transdermal nitroglycerin in stable angina pectoris. *Eur Heart J* 1988;9(Suppl A):73–81.
49. Greco R, Schiattarella M, Wolff S, et al. Long-term efficacy of nitroglycerin patch in stable angina pectoris. *Am J Cardiol* 1990;65:9J–15J.
50. Bassan M. The day-long antianginal effectiveness of nitroglycerin patches. *Chest* 1991;99:1120–1125.
51. Scardi S, Camareni F, Pandullo G, et al. Efficacy of continuous and intermittent transdermal treatment with nitroglycerin in effort induced angina pectoris: A multicenter study. *Int J Cardiol* 1991;32:241–248.
52. Steering Committee, Transdermal Nitroglycerin Cooperative study. Acute and chronic antianginal efficacy of continuous twenty-four-hour application of transdermal nitroglycerin. *Am J Cardiol* 1991;68:1263–1273.
53. Klemsdal TO, Gjesdal K. The effect of transdermal nitroglycerin on exercise tolerance in relation to patch application time—a meta-analysis. *Cardiovasc Drugs Ther* 1992;6:641–649.
54. Bennett D, Daviesa A, Davis A. Sustained anti-anginal action of glyceryl trinitrate cream. *Br J Clin Pharmacol* 1983;15:173–180.
55. Parker JO, Van Koughnett KA, Fung HL. Transdermal isosorbide dinitrate in angina pectoris. *Am J Cardiol* 1984;54:8–13.
56. Handler CI, Sullivan ID. Double-blind randomized cross-over trial comparing isosorbide dinitrate cream and oral sustained-release tablets in patients with angina pectoris. *Int J Cardiol* 1985;7:149–156.
57. Thadani U. The effectiveness of transcutaneous nitrate preparations for angina pectoris. *Int J Cardiol* 1987;14:9–14.
58. Abrams J. The brief saga of transdermal nitroglycerin discs: Paradise lost? *Am J Cardiol* 1984;64:220–224.
59. Colditz GA, Halvorsen KT, Goldhaber SZ. Randomized clinical trials of transdermal nitroglycerin systems for the treatment of angina: A meta analysis. *Am Heart J* 1988;116:174–180.
60. Abrams J. Transcutaneous nitroglycerin—ointment or disc? *Am Heart J* 1984;108:1597–1600.
61. Bogaert MA, Deschaepdryver AF. Tolerance towards glyceryl trinitrate (trinitrin) in dogs. *Arch Int Pharmacodyn Ther* 1968;171:221–224.
62. Needleman P. Tolerance to the vascular effects of glyceryl trinitrin. *J Pharmacol Exp Ther* 1970;171:98–102.

63. Needleman P, Johnson EM Jr. Mechanism of tolerance to organic nitrates. *J Pharmacol Exp Ther* 1973;184:709–715.
64. Bassange E. Nitrates in differnt vascular beds: Nitrate tolerance and interaction with endothelial function. *Am J Cardiol* 1992;70:23B–29B.
65. Fung HL, Chong S, Kowaulk E, et al. Mechanism for pharmacologic interaction of nitrates with thiols: Existence of an extracellular pathway for the reversal of nitrates by n-acetylcysteine. *J Pharmacol Exp Ther* 1988:524–530.
66. Schwartz AM. The cause, relief and prevention of headaches arising from contact with dynamite. *N Engl J Med* 1946; 235:651.
67. Morton WE. Occupational habituation to aliphatic nitrates and the withdrawal hazards of coronary disease and hypertension. *J Occup Med* 1977;19:197–199.
68. Klock JC. Nonocclusive coronary disease after chronic exposure to nitrates for physiologic nitrate dependence. *Am Heart J* 1975;89:510.
69. Thadani U, Whitsett T. Relationship of pharmacokinetics and pharmacodynamic properties of organic nitrates. *Clin Pharmacokinet* 1988;15:32–43.
70. Thadani U, Whitsett T, Hamilton SF. Nitrate therapy for myocardial ischemic syndrome: Current perspectives including tolerance. *Curr Probl Cardiol* 1988;13:725–784.
71. Gjesdal K, Klemsdal TO, Rykke EO, et al. Transdermal nitrate therapy: Bioavailability during exercise increases transiently after the daily change of patch. *Br J Clin Pharmacol* 1991;31:560–562.
72. Zimrin D, Reichek N, Bogin K, et al. Antianginal effects of intravenous nitroglycerin over 24 hours. *Circulation* 1988;77:1376–1384.
73. Cowan JC, Bourke JP, Reid DS. Prevention of tolerance to nitroglycerin patches by overnight removal. *Am J Cardiol* 1987;60:271–275.
74. Luke R, Sharpe N, Coxon R. Transdermal nitroglycerin in angina pectoris: Efficacy of intermittent application. *J Am Coll Cardiol* 1987;10:642–646.
75. Nabel EG, Barry J, Rocco MB, et al. Effects of dosing intervals on the development of tolerance to high dose transdermal nitroglycerin. *Am J Cardiol* 1989;63:664–665.
76. Waters DD, Juneau M, Gossard D, et al. Limited usefulness of intermittent nitroglycerin patches in stable angina. *J Am Coll Cardiol* 1989;13:421–425.
77. de Milliano PA, Koster RW, Bar FW. Long-term efficacy of continuous and intermittent use of transdermal nitroglycerin in stable angina pectoris. *Am J Cardiol* 1991;68:857–862.
78. James MA, Papouchado M, Jones JV. Attenuation of nitrate effect during an intermittent treatment regimen and the time course of nitrate tolerance. *Eur Heart J* 1991;12:1266–1272.
79. Knuuti MJ, Wahl M, Wiklund I, et al. Acute and long-term effects on myocardial ischemia of intermittent and continuous transdermal nitrate therapy in stable angina. *Am J Cardiol* 1992;69:1525–1532.
80. Scardi S, Camerini F, Pandullo C, et al. Efficacy of continuous and intermittent transdermal treatment with nitroglycerin in effort angina pectoris: A multicenter study. *Int J Cardiol* 191;32:241–248.
81. Schaer DH, Buff LA, Katz RJ. Sustained antianginal efficacy of transdermal nitroglycerin patches using an overnight 10-hour nitrate free interval. *Am J Cardiol* 1988;61:46–50.
82. DeMots H, Glasser SP and the Transderm-Nitro Study Group. Intermittent transdermal nitroglycerin therapy in the treatment of chronic stable angina. *J Am Cardiol* 1989;13:786–793.

83. Fox KM, Dargie HJ, Deanfield J, et al. Avoidance of tolerance and lack of rebound with intermittent dose titrated transdermal glyceryl trinitrate. *Br Heart J* 1991;66:151–155.
84. Rossett E, Luca C, Bonetti F, et al. Transdermal nitroglycerin reduces the frequency of anginal attacks but fails to prevent silent ischemia. *J Am Coll Cardiol* 1993;21:337–342.
85. Holdright DR, Katz RJ, Wright CA, et al. Lack of rebound during intermittent transdermal treatment with glyceryl trinitrate in patients with stable angina on background β blocker. *Br Heart J* 1993;69:223–227.
86. Mulcahy D, Keegan J, Cunningham D, et al. Circadian variation of total ischemic burden and its alteration with antianginal agents. *Lancet* 1989;2:755–758.
87. Thadani U. Medical therapy of stable angina pectoris. *Cardiol Clin* 1991;9:73–85.
88. Thadani U. Role of nitrates in angina pectoris. *Am J Cardiol* 1992;70:43B–53B.

Nitrates for Unstable Angina

Udho Thadani and Lionel H. Opie

Summary. The term *unstable angina* encompasses heterogeneous clinical syndromes. Fissuring of an atherosclerotic coronary artery plaque with superimposed platelet deposition, with or without additional thrombus formation, is invariably responsible for a prolonged episode of angina at rest, increasing frequency of angina at rest, or with minimal exertion of less than 4 weeks in duration and early postinfarction angina. Plaque progression, rather than plaque fissuring, is the most likely mecahnism for progressive reduction in walking distance due to angina in patients who previously have stable angina. Coronary artery spasm is responsible for Prinzmetal's variant angina, but its exact role in other forms of unstable angina is unknown. The mainstay of treatment of unstable angina (prolonged episode of angina at rest and recent onset angina at rest, or with minimal exertion with a crescendo pattern) is aspirin, heparin, or both. Both aspirin and intravenous (IV) heparin or their combination reduce early mortality and the incidence of acute myocardial infarction in patients hospitalized with unstable angina. However, these agents do not promptly relieve chest pain. There are no placebo-controlled studies evaluating the usefulness of nitrates in unstable angina. In open-label studies, continuous therapy with IV nitroglycerin (NTG) for 24 hours or longer has been shown to relieve chest pain in patients with rest angina refractory to therapy with other antianginal agents, including long-acting nitrates. Recurrence of chest pain in patients receiving IV NTG is a common problem and probably represents development of pharmacologic tolerance, but this can be overriden by dose escalation; protracted tolerance during short-term use of IV NTG is usually not a problem. In the acute phase of unstable angina, IV NTG is the preparation of choice as the dose can be rapidly titrated up or down. There is no role of intermittent nitrate therapy in the acute phase of unstable angina. Once the patient is stable for 12–24 hours, IV NTG should be tapered gradually and intermittent therapy with a long-acting nitrate, as outlined for the treatment of stable angina, instituted. Aspirin reduces mortality and morbidity during long-term therapy and should be continued indefinitely. Routine use of morphine and other potent analgesics is not recommended. Patients who do not respond to IV NTG or in whom IV NTG is contraindicated should be treated with a beta-blocker devoid of intrinsic sympathomimetic activity, provided there are no contraindications to beta-blocker therapy. The role of calcium channel blockers in patients nonresponsive

to IV NTG is less well defined. In patients already receiving beta-blockers and nitrates, the addition of nifedipine may be beneficial. However, monotherapy with nifedipine or other first-generation dihydropyridines is not recommended. Although there are no large trials of diltiazem or verapamil in unstable angina, these agents are often used in patients who are not candidates for beta-blocker therapy. Patients who are refractory to intensive medical therapy are candidates for coronary angiography and revascularization procedures, provided the coronary anatomy is suitable for such procedures.

Key Words. **unstable angina, nitrates, nitrate tolerance, nitrate dose escalation to override tolerance**

The term *unstable angina* has been used to describe a variety of heterogeneous clinical syndromes (Table 1), ranging from a prolonged episode of angina at rest, recent onset (within 4 weeks) rest, or exertional angina with progressive deterioration, to worsening of exercise tolerance in patients with stable angina, postinfarction angina, and recurrence of angina following angioplasty. A comprehensive subclassification was recently proposed [1]. Patients with preexistent stable angina who experience progressive deterioration in their exercise performance and those with recent onset exertional angina usually have progression of underlying coronary disease [2,3]. Prolonged episode of angina at rest of more than 15 minutes and progressive angina at rest is invariably due to fissuring or rupture of a coronary artery plaque, which previously might have produced no symptoms [4–9]. The denuded endothelium predisposes to platelet aggregation and thrombus formation [10–12]. Deposition of platelets reduces ostial lumen with a resultant reduction in blood flow. The same pathophysiologic process is responsible for recurrence of angina at rest, which some patients experience for the first few days after an acute myocardial infarction [3]. However, after 7–10 days postinfarction, exertional angina is usually due to residual coronary stenosis in the infarct related artery, which limits blood flow to the non-necrotic myocardium during exertion. Coronary artery spasm is primarily responsible for Prinzmetal's variant angina [3]; its role in other forms of unstable angina is poorly defined.

The mainstay of treatment of unstable angina (prolonged episode of angina at rest, progressive recurrent episodes of angina at rest or with minimal

Table 1. *Clinical syndromes of unstable angina*[a]

1. Recent onset of angina at rest or on exertion with a crescendo pattern (preinfarction angina), without provocation, within the past 4 weeks.
2. Prolonged episode of angina at rest lasting more than 15 minutes.
3. Worsening exertional angina in a patient with stable angina.
4. Recent-onset exertional angina without a crescendo pattern.
5. Angina of recent onset post acute myocardial infarction.
6. Angina post coronary angioplasty.
7. Prinzmetal's variant angina.

[a] For further subclassification, see Braunwald [1]. Only syndromes under category 1 and 2 are considered true forms of unstable angina and are the focus of this review.

exertion of less than 4 weeks duration) is aspirin, heparin, or both [13–18]. Further discussion pertains to these forms of unstable angina.

In patients hospitalized with unstable angina, aspirin therapy consistently reduces early morbidity and mortality [15,16]. Further long-term therapy with aspirin also reduces the incidence of myocardial infarction and death compared to placebo [13,14,18]. Intravenous heparin during the acute phase of unstable angina is either as effective or more effective than aspirin alone, and the combination of low-dose (75 mg) aspirin and IV heparin also reduces the incidence of refractory angina and the need for revascularization [15,17].

Neither aspirin nor heparin promptly relieve chest pain or reduce myocardial ischemia. Nitrates, beta-blockers, and calcium channel blockers are widely used to treat chest pain and myocardial ischemia in patients with unstable angina, but monotherapy with the calcium channel blocker, nifedipine, has been associated with an increase in mortality and the incidence of myocardial infarction [19]. Low-dose IV nitroglycerin (NTG) therapy in acute myocardial infarction reduces mortality and infarct-related complications [20,21]. Whether nitrates improve survival or reduce the incidence of myocardial infarction in patients with unstable angina is, however, unknown. Intravenous NTG can, however, rapidly relieve angina and myocardial ischemia, even in patients refractory to other antianginal agents, including oral or topical nitrates [22]. This review deals with the rationale of nitrate therapy in unstable angina (prolonged episode of angina at rest, progressive recurrent episodes of angina at rest, or minimal exertion) and provides guidelines for the use of nitrates on the basis of published literature.

Mechanism of Action of Nitrates in Unstable Angina

Possible beneficial effects of initial administration of nitrates in unstable angina are summarized in Table 2 [23–30]. Nitrates improve the balance between myocardial oxygen supply and demand. This is achieved by their predominant venodilator effect with a resultant decrease in venous return and ventricular volumes and end-diastolic pressures. In addition, nitrates reduce afterload and produce dilation of epicardial arteries, collateral channels, and stenotic sites. It is also conceivable that dilation of a stenotic site induced by nitrates might permit downstream washing of platelet aggregates at the site of plaque rupture with resultant relief of chest pain and myocardial ischemia. Nitrates have been shown to prevent platelet aggregation in vitro [27,29,30] and to reduce platelet deposition and vasoconstriction at the site of experimental endothelial injury [27]. Whether this effect of nitrates on platelets provides an added benefit in patients receiving aspirin is unknown but is probable, as the nitrate effect is mediated via nitric oxide [31], while the aspirin effect is mediated via the cyclooxygenase pathway [32]. In summary, nitrates have a dual beneficial effect, as these agents increase blood flow to the ischemic area and reduce myocardial oxygen demand.

Do Nitrates Reduce the Incidence of Myocardial Infarction and Mortality in Unstable Angina?

In patients with acute myocardial infarction, especially anterior myocardial infarction, cautious use of IV NTG was associated with a reduced mortality and infarct-related complications compared with placebo [20]. Beneficial effects of nitrates on mortality were also reported in a meta-analysis of the published data prior to 1988 [21]. However, in the ISIS IV study presented at the AHA meeting in Atlanta in November 1993, treatment with imdur 60 mg once a day was not superior to placebo. There are no data showing that nitrates reduce mortality or lower the incidence of myocardial infarction in patients with unstable angina [22].

Use of Nitrates in the Acute Phase of Unstable Angina: Effects on Chest Pain and Myocardial Ischemia

In previous studies in unstable angina, none of which were placebo controlled, nitrates were used in addi-

Table 2. *Mechanism of action of intravenous nitroglycerin in unstable angina*

Decreased myocardial oxygen demand	Enhanced myocardial oxygen supply
1. Decreased left ventricular end-diastolic volume (preload) decreases wall stress. 2. Decreased cardiac output decreases arterial pressure. 3. Decreased arterial compliance and peripheral resistance lowers systolic pressure (afterload).	1. Dilation of coronary arteries: epicardial vessels, eccenteric stenosis, and collaterals. 2. Decreased left ventricular end-diastolic pressure enhances subendocardial flow. 3. Inhibition of coronary spasm or reduction of coronary vasomotor tone. 4. Decreased platelet aggregation and deposition prevents coronary artery occlusion and vasospasm (?). 5. Dilation of epicardial vessel at the site of culprit lesion facilitates dislodgement of occlusive platelet plug (?).

Adapted from Thadani [22], with permission.

tion to beta-blockers and/or calcium channel blockers, but aspirin or heparin was not employed [32–42]. The majority of the patients in these studies required surgical intervention, suggesting that therapy with nitrates in the absence of concomitant antiplatelet therapy did not reduce the need for revascularization.

Thus, the usefulness of nitrates in patients with unstable angina remains unproven. In the only published placebo-controlled study entitled "Management of Vasospastic Angina at Rest with Continuous Infusion of Isosorbide Dinitrate [43]," patients had variant rather than unstable angina. In these patients long-term therapy with nitrates was effective. However, these patients experienced angina at rest due to coronary artery spasm rather than as a sequel to plaque rupture and platelet deposition. Although therapy with nitrates is effective in variant angina, calcium channel blockers, including nifedipine, are the treatment of choice.

Topical nitrates, especially NTG ointment, are used by some physicians in the acute phase of unstable angina. However, continued efficacy of ointment in unstable angina remains to be documented.

In the published studies, the effects of nitrates on the number of episodes of angina at rest were compared to baseline data obtained at the time of hospitalization [32–42]. This approach, however, may provide erroneous information, as the majority of the patients with unstable angina often become pain free over time after they have received antiplatelet therapy. In the absence of a parallel group treated with placebo, it would be impossible to ascertain whether nitrates are truly beneficial in unstable angina.

The published data (Table 3) derived from openlabel studies, however, support the prevailing clinical opinion that IV NTG is useful for the rapid relief of angina at rest and myocardial ischemia, and its use is associated with a reduced need for morphine analgesia in patients who are nonresponsive to high doses of beta-blockers, oral and topical nitrates, and/or calcium channel blockers [32–42]. Intravenous NTG for 24–72 hours has been used in the initial management of patients who have angina at rest [22]. Either complete relief or consistent reduction in the number of anginal attacks and ischemic episodes has been reported in 8 of the 10 published studies [22] (Table 3). It should, however, be recognized that, with the exception of one study [42], aspirin, heparin, or both were not routinely used.

Table 3. *Studies with intravenous nitroglycerin (NTG) in unstable angina*[a]

Reference	n	NTG dose (µg/min) mean (range)	Reduction (complete relief) of anginal episodes	Comments
Dauwe et al. [33]	14	47 (30–180)	100	Refractory angina despite beta-blocker and long-acting nitrate therapy
Page et al. [34]	17	40	98	Refractory to beta-blocker and oral nitrates
Brodsky et al. [35]	14	13 (30–350)	36	Refractory angina despite long-acting nitrate therapy
				Hypotension a major problem during intravenous nitroglycerin
Mikolich et al. [36]	45	55 (5–267)	89 (89)	Refractory to beta-blocker and multiple doses of sublingual nitroglycerin
DePace et al. [37]	20	72 (15–225)	85	Refractory angina despite beta-blocker, oral nitrate, and nifedipine therapy
Roubin et al. [38]	16	13	100 (38)	Refractory angina despite beta-blocker, nitrate, and calcium blocker therapy
Squire et al. [39]	42	(277–393)	45	Refractory angina despite beta-blocker and long-acting nitrate therapy
Curfman et al. [40]	40	82 (100–200)	70 (36)	Background beta-blocker therapy: fixed- dose nitrate therapy during the study
Kaplan et al. [41]	35	140 (50–350)	94 (71)	Refractory angina despite beta-blocker and long-acting nitrate therapy
Thadani et al. [42]	43	(10–960)	98—(32)	Tolerance development necessitated increase of NTG dose

[a] In none of the studies, with the exception of the study by Thadani et al., were aspirin and/or heparin therapy used concomitantly. Adapted with permission from Conti [44] and Thadani [22].

Therapy with IV NTG can be rationally recommended in patients who continue to have angina at rest despite therapy with aspirin and/or IV heparin and sublingual NTG. In patients who are pain free at the time of hospitalization and those who respond to sublingual NTG, it is unknown whether IV NTG will prevent the recurrence of myocardial ischemia and chest pain at rest. Placebo-controlled studies are needed, prior to recommending the routine use of IV NTG, in this group of patients.

Routine use of morphine analgesia cannot be recommended as the awareness of the recurrence of ischemic chest pain, which may be a warning signal of an impending infarction, may be masked by a potent analgesic.

Comparison of Oral, Topical, and Intravenous NTG in Unstable Angina

Curfman et al. [40] evaluated the usefulness of high doses of oral isosorbide dinitrate plus NTG ointment with that of IV NTG therapy. Both were found to be effective, but more patients experienced relief of angina during IV NTG therapy. This study, however, was unusual in that fixed doses of nitrates were used. The dose was chosen on the basis of reduction in blood pressure of 15 mmHg and dose escalation was not permitted. There are no other studies evaluating different nitrate formulations in unstable angina.

Does tolerance develop during continuous intravenous NTG therapy in unstable angina?

Continuous or frequent intermittent therapy with any form of nitrate preparations leads to the development of tolerance [45–46]. In patients with stable angina pectoris, tolerance developed within 24 hours of high-dose continuous IV NTG therapy [46]. The study of nitrate tolerance in unstable angina is fraught with problems due to underlying pathophysiologic processes responsible for unstable angina. One is never certain whether recurrence of anginal pain while patients are receiving IV NTG is due to the development of pharmacologic tolerance or due to intermittent platelet deposition or thrombus formation.

In an open-label study in unstable angina, tolerance as defined by recurrence of angina at rest during IV NTG therapy developed in 65% of patients, but it was overridden by increasing the dose of NTG, which promptly relieved ischemic pain [42]. In that study, 43 hospitalized patients with unstable angina were evaluated. Despite treatment with other antianginal agents, including long-acting nitrates as well as aspirin, heparin, or both, these patients still had angina at rest. They were initially treated with IV NTG, with a starting dose of 5 μg per minute, which was increased progressively until relief of angina or a fall in systolic blood pressure to below 90 mmHg. Thereafter, the infusion was continued for up to 36 hours. One patient experienced no relief of pain, despite receiving more than 200 μg/minute of IV NTG. Fourteen of the 43 patients had no recurrence of angina, and NTG was tapered off and discontinued after 7–24 hours without recurrence of pain. Chest pain recurred in the remaining 28 patients within an average of 8.5 hours (range 1–20 hours), which responded to higher doses of NTG (40–240 μg/min). Further recurrence of pain within an average of 3 hours (range 1–7 hours) in 7 of 28 patients necessitated a further increase in the dose of IV NTG. Subsequently, the majority of the patients were weaned from IV NTG without problems and placed on oral intermittent nitrate therapy.

This and other reports that IV NTG relieves chest pain and myocardial ischemia in patients already receiving high doses of oral nitrates suggests that tolerance to nitrates can be overridden by dose escalation of IV NTG for the short term. These observations cast doubt on the importance of sulfhydryl donor depletion as the cause of nitrate tolerance.

Horowitz reported that tolerance to IV NTG can be prevented by concomitant use of N-acetylcysteine [47]. However, it is more likely that N-acetylcysteine potentiated nitrate effect rather than reversed tolerance. He recommended low-dose IV NTG therapy in order to avoid protracted nitrate tolerance [48]. Available data from open-label studies, however, do not support his view. Progressive increase in the dose of IV NTG rapidly relieves recurrent anginal pain [22,33–42]. Furthermore, high-dose therapy with IV NTG can usually be weaned off after 24–72 hours without a clinical rebound in the majority of patients [42].

Justification of Use of Intravenous NTG Over Topical and Oral Nitrates

During the acute phase of unstable angina, one is uncertain whether the patient will stabilize or will eventually develop a myocardial infarction. Given this uncertainty, and unexpected rapid changes in the clinical state of these patients, the nitrate preparation of choice would be the one that could be rapidly discontinued in case the patient develops hypotension or be given in higher quantities if angina recurs, or hypertension or signs of left ventricular failure develop.

Only IV NTG fulfills the criteria discussed above. Topical nitroglycerin ointment can be removed rapidly, but upward dose titration is difficult. From the available data, IV NTG is a formulation of choice; it is effective when other forms of nitrates have been ineffective [22]. It permits immediate delivery and ease of titration. Intravenous isosorbide dinitrate is used in Europe and is claimed to be efficacious in unstable angina [49], but it has a prolonged action, which may pose problems if an adverse reaction were to develop.

Guidelines for Intravenous Nitroglycerin Therapy

Patients with unstable angina, who continue to have resting chest pain despite therapy with sublingual NTG, should be treated with IV NTG provided the patient does not have symptomatic hypotension or a systolic blood pressure less than 95 mmHg. The starting dose should be 5–10 μg/min and the dose can be titrated upwards every 5–10 minutes until relief of chest pain or a fall in systolic blood pressure to 90 mmHg has occurred, or development of adverse effects necessitates reduction of dosage [22,42]. If anginal pain recurs, the dose should be titrated upwards as often as necessary until pain is relieved. Doses as high as 1000 μg/min may be required, but if adverse effects occur discontinuation of therapy is warranted.

Intravenous NTG can be used for up to 72 hours or longer, but attempts should be made to taper the drug gradually after 24–48 hours of therapy. If the patient does not experience a recurrence of angina, conventional antianginal therapy, including intermittent use of nitrates, should be instituted. In patients who became pain free after sublingual NTG, the role of IV NTG is unclear. However, if the patient experiences recurrence of angina at rest, therapy with IV nitroglycerin, as discussed above, should be instituted.

Is there any Role of Intermittent Nitrate Therapy in Unstable Angina?

Recent recommendations of nitrate use in stable angina are to use these agents only during the day and to provide nitrate-free periods at night and during the early hours of the morning [50,51]. However, this strategy cannot be recommended in the acute phase of unstable angina. These patients are at risk of developing an acute myocardial infarction or even sudden death. A rebound increase in angina at night during nitrate low or free periods during intermittent therapy reported in patients with stable angina remains a major concern in patients with unstable angina, as this would further complicate management of these patients [51]. Therefore, only continuous therapy with IV NTG for the first 24–72 hours is recommended. It should be recognized that nitrate tolerance does develop, even during short-term therapy with IV NTG, but it can be circumvented by dose escalation [42].

Interaction Between Intravenous NTG and Heparin

Concomitant use of IV NTG and heparin may reduce therapeutic efficacy of heparin [52,53]. Therefore, partial prothrombin time should be closely monitored in patients who are receiving these two drugs simultaneously. The dose of heparin may need to be adjusted upwards during continuous use of IV NTG therapy and downwards when NTG therapy is withdrawn.

Combination Therapy of Intravenous NTG and Beta-Blocker Therapy or Calcium Channel Blockers

No controlled studies have been performed evaluating the combination of IV NTG and beta-blockers or calcium channel blockers in unstable angina. Beta-blockers are effective antianginal and antiischemic agents and have been shown to reduce the incidence of myocardial infarction in patients with unstable angina and those hospitalized with threatened infarction [19,54]. However, it is not known if these agents reduce morbidity and mortality in patients who are also receiving aspirin and/or heparin therapy. Beta-blockers are often used in patients with unstable angina who remain symptomatic despite therapy with nitrates. Addition of a beta-blocker (nonselective or cardioselective) often relieves chest pain and myocardial ischemia [55]. However, beta-blockers with intrinsic sympathomimetic activity, such as pindolol, should be avoided, as these agents may aggravate anginal episodes at rest [55].

There are no large trials with calcium channel blockers in unstable angina. In a placebo-controlled study, nifedipine given in addition to nitrates increased cardiovascular morbidity and mortality [19]. However, when nifedipine was given to patients already on beta-blockers and nitrates, there was a reduction in morbidity compared to placebo [19]. The usefulness of slow-release formulations of nifedipine and second- and third-generation dihyropydridine agents, including amlodipine, in unstable angina is unknown. The effects of diltiazem or verapamil, when given in addition to IV NTG, on cardiovascular morbidity and mortality in unstable angina are unknown, although these agents are often used in patients refractory to IV NTG or who are not candidates for beta-blocker therapy.

Failed Nitrate Therapy

Patients who remain symptomatic despite therapy with aspirin, heparin, IV nitroglycerin, and beta-blockers and/or calcium channel blockers may be at a greater risk for future cardiac events, and in these patients coronary angiography should be performed. Those with poor left ventricular function and triplevessel coronary disease should undergo bypass surgery. Patients with one- or two-vessel disease should be considered for either balloon angioplasty or bypass surgery. However, angioplasty of the culprit lesion, as well as surgery, carry an increased risk.

Adverse Effects of Intravenous Nitroglycerin

Excessive hypotension may develop during initiation of IV NTG therapy and would necessitate discontinuation of infusion. Headaches are common, but invariably respond to acetaminophen, but if unresponsive to analgesics IV NTG should be discontinued. Methemoglobinemia is usually not a problem despite therapy with high doses of IV NTG for 24–72 hours, but should be watched for in patients who have concomitant severe obstructive airway disease.

Nitrates for the Late or Stable Phase of Unstable Angina

The first 24–72 hours are critical in the management of patients with unstable angina. With appropriate management, the majority of these patients can be treated medically [56]. Aspirin therapy should be continued indefinitely and IV heparin can usually be discontinued after 24–72 hours. Intravenous NTG should be gradually tapered rather than abruptly stopped for fear of a rebound increase in angina. Before 1–2 hours of stopping IV NTG, patients should be started on an oral or a topical nitrate preparation. In order to avoid development of tolerance, intermittent nitrate therapy is recommended. Thus, there is always a risk that patients may experience nocturnal angina during the nitrate-free or low-nitrate periods at night and during the early hours of the morning. Therefore, it is preferable to use a beta-blocker concomitantly with intermittent nitrate therapy to provide 24 hour antianginal prophylaxis [50]. If there are contraindications to beta-blocker therapy, diltiazem or verapamil, rather than a 4-dihydropyridine agent, should be used concomitantly with nitrates [57].

Prior to hospital discharge, patients can be stratified into low- and high-risk groups by performing an exercise test with or without gated blood pool studies to evaluate left ventricular function. Alternatively, thallium perfusion studies can be performed to assess the size of myocardium at jeopardy. Management of these patients, however, is not different to those with stable angina pectoris.

An alternative to risk stratification by noninvasive tests is routine coronary angiography, which identifies patients with left main disease and those with normal coronary arteries. However, cardiac catheterization is an invasive procedure and cannot be routinely recommended.

References

1. Braunwald E. Unstable angina: A classification. *Circulation* 1989;80:410–414.
2. Moise A, Théroux P, Taeymans Y, et al. Unstable angina and progression of coronary atherosclerosis. *N Engl J Med* 1983;309:685–689.
3. Théroux P, Lidón RM. Unstable angina: Pathogenesis, diagnosis and treatment. *Curr Probl Cardiol* 1993;18:157–232.
4. Davies MJ, Thomas AC. Plaque fissuring—the cause of acute myocardial infarction, sudden ischemic death, and crescendo angina. *Br Heart J* 1985;53:363–373.
5. Falk E. Plaque rupture with severe preexisting stenosis precipitating coronary thrombosis: Characteristics of coronary atherosclerotic plaques underlying fatal occlusive thrombi. *Br Heart J* 1983;50:127–134.
6. Falk E. Unstable angina with fatal outcome: Dynamic coronary thrombi leading to infarction and/or sudden death: Autopsy evidence of recurrent mural thrombosis with peripheral embolization culminating in total vascular occlusion. *Circulation* 1985;71:699–708.
7. Fuster V, Chesebro JH. Mechanisms of unstable angina. *N Engl J Med* 1986;315:1023–1025.
8. Ambrose JA, Winters SL, Stern A, et al. Angiographic morphology and the pathogenesis of unstable angina. *J Am Coll Cardiol* 1985;5:609–616.
9. Fuster V, Badimon L, Badimon JJ, et al. The pathogenesis of coronary artery disease and the acute coronary syndrome. *N Engl J Med* 1992;326:310–318.
10. Fitzgerald DJ, Roy L, Catella F, et al. Platelet activation in unstable coronary disease. *N Engl J Med* 1986;315:983–989.

11. Hamm CW, Lorenz RL, Bleifield W, et al. Biochemical evidence of platelet activation in patients with persistent unstable angina. *J Am Coll Cardiol* 1987;10:998–1004.
12. Théroux P, Latour JG, Delara J, Léger-Gauthier C. Fibrino peptide A and platelet factor levels in unstable angina. *Circulation* 1987;75:156–162.
13. Lewis HD Jr, Davis JW, Archibald DG, et al. Protective effects of aspirin against acute myocardial infarction and death in men with unstable angina: Results of a Veterans Administration Cooperative Study. *N Engl J Med* 1983;309:396–403.
14. Cairns JA, Gent M, Singer J, et al. Aspirin, sulfinpyrazone or both in unstable angina: Results of a Canadian multicenter trial. *N Engl J Med* 1985;313:1369–1375.
15. Théroux P, Ouimet H, McCans J, et al. Aspirin, heparin or both to treat acute unstable angina. *N Engl J Med* 1988;319:1105–1111.
16. Théroux P, Waters D, Lam J, et al. Reactivation of unstable angina following discontinuation of heparin. *N Engl J Med* 1992;327:141–145.
17. The Risc Group. Risk of myocardial infarction and death during treatment with low dose aspirin and intravenous heparin in men with unstable coronary artery disease. *Lancet* 1990;336:827–830.
18. Wallentin LC, and the Research Group on Instability in Coronary Artery Disease in Southeast Sweden. Aspirin (75 mg/day) after an episode of unstable coronary artery disease. Long-term effect on the risk of myocardial infarction, occurrence of severe angina and the need for revascularization. *J Am Coll Cardiol* 1991;18:1587–1593.
19. Holland Interuniversity Nifedipine/Metoprolol Trial (HINT) Research Group. Early treatment of unstable angina in the coronary care unit: A randomized double-blind, placebo controlled comparison of recurrent ischemia in patients treated with nifedipine or metoprolol or both. *Br Heart J* 1986;56:400–413.
20. Jugdutt BI, Warnica JW. Intravenous nitroglycerin therapy to limit myocardial infarct size, expansion and complications; effect of timing, dosage and infarct size. *Circulation* 1988;78:906–919.
21. Yusuf S, Collins R, MacMohan S, Peto R. Effect of intravenous nitrates on mortality in acute myocardial infarction: An overview of the radomized trials. *Lancet* 1988;1:1088–1092.
22. Thadani U. Nitrate therapy in unstable angina. In: Bleifield W, Hamm CW, Braunwald E, eds. *Unstable Angina*. Berlin: Springer-Verlag 1990:203–213.
23. McGregor M. Pathogenesis of angina pectoris and role of nitrates in relief of myocardial ischemia. *Am J Med* 1983;74(Suppl 4):21–27.
24. Brown BG, Bolson E, Peterson RB, et al. The mechanism of nitroglycerin action: Stenosis vasodilation as a major component of dosing response. *Circulation* 1981;64:1089–1097.
25. Winbury MM, Howe BB, Hefner MA. Effect of nitrates and other coronary dilators on large and smaller coronary vessels: A hypothesis for the mechanism of action of nitrates. *J Pharmacol Exp Ther* 1969;168:70–95.
26. Winbury WM, Howe BB, Weiss HR. Effect of nitroglycerin and dipyridamole on epicardial and endocardial oxygen tension: Further evidence for redistribution of myocardial flow. *J Pharmacol Exp Ther* 1971;176:184–199.
27. Lam JYT, Chesebro JH, Fuster V. Platelets, vasoconstriction and nitroglycerin during arterial wall injury: A new antithrombotic role of an old drug. *Circulation* 1988;78:712–716.

28. Levin RI, Jaffe EA, Weksler BB, et al. Nitroglycerin stimulates synthesis of prostacyclin by cultured human endothelial cells. *J Clin Invest* 1981;67:762–769.
29. Folts JD, Stamler J, Loscalzo J. Intravenous nitroglycerin infusion inhibits cyclic blood flow response caused by periodic platelet thrombus formation in stenosed coronary arteries. *Circulation* 1991;83:2122–2127.
30. Stamler JS, Loscalzo J. The antiplatelet effects of organic nitrates and related nitroso compounds in vitro and in vivo and their relevance in cardiovascular disorders. *J Am Coll Cardiol* 1991;18:1529–1536.
31. Palmer RMJ, Ferrige AC, Moncada S. Nitric oxide release accounts for the biological activity of endothelial derived relaxing factor. *Nature* 1987;327:524–526.
32. Moncada S, Vane JP. Arachidonic acid metabolites and the interaction between platelets and blood-vessel wall. *N Engl J Med* 1979;300:1142–1147.
33. Dauwe F, Affaki G, Waters DD, et al. Intravenous nitroglycerin in refractory unstable angina (abstr). *Am J Cardiol* 1979;43:416.
34. Page A, Hurtle JF, Roudaut R, et al. Intravenous trinitrin in the treatment of pre-infarctional angina. *Nouve Presse Med* 1979;8:268–270.
35. Bordsky SJ, Halperin JL, Klein MD, et al. Intravenous nitroglycerin infusion in unstable angina (abstr). *Clin Res* 1980;28:608A.
36. Mikolich RJ, Nicoloff NB, Robinson PH, et al. Relief of refractory angina with continuous intravenous infusion of nitroglycerin. *Chest* 1980;77:375–379.
37. DePace ML, Hering IM, Kotler MN, et al. Intravenous nitroglycerin for rest angina. Potential pathophysiologic mechanisms of action. *Arch Intern Med* 1982;142:1806–1809.
38. Roubin GS, Harris PJ, Eckhardt L, et al. Intravenous nitroglycerin in refractory unstable angina pectoris. *Aus NZ J Med* 1982;12:598–602.
39. Squire A, Cantor R, Packer M. Limitations of continuous intravenous nitroglycerin prophylaxis in patients with refractory angina at rest (abstr). *Circulation* 1982;66(Suppl II):II120.
40. Curfman GD, Heinsimer JA, Lozner EC, et al. Intravenous nitroglycerin in the treatment of spontaneous angina pectoris: A prospective, randomized trial. *Circulation* 1983;67:276–282.
41. Kaplan K, Davison R, Parker M, et al. Intravenous nitroglycerin for the treatment of angina at rest unresponsive to standard nitrate therapy. *Am J Cardiol* 1983;51:694–698.
42. Thadani U, Sharma M, Thompson D, et al. Intravenous nitroglycerin and tolerance in unstable angina (abstr). *Circulation* 1987;76(Suppl IV):IV128.
43. Disdante A, Maseri A, Severi S, et al. Management of vasospastic angina at rest with continuous infusion of isosorbide dinitrate. *Am J Cardiol* 1979;44:533–539.
44. Conti R. Use of nitrates in unstable angina pectors. *Am J Cardiol* 1987;60:31H–34H.
45. Thadani U, Whitsett T, Hamilton SF. Nitrate therapy for myocardial ischemic syndrome: Current perspectives including tolerance. *Curr Probl Cardiol* 1988;13:725–784.
46. Zimrin D, Reichek M, Bogin KT, et al. Antianginal effects of intravenous nitroglycerin over 24 hours. *Circulation* 1988;77:1376–1384.
47. Horowitz JD, Henry LA, Syrjanen ML, et al. Combined use of nitroglycerin and N-acetylcysteine in the management of unstable angina pectoris. *Circulation* 1988;77:787–794.

48. Horowitz JD. Role of nitrates in unstable angina pectoris. *Am J Cardiol* 1992;70(Suppl):64B–71B.
49. Weber S, Auclert L, Touizak, et al. Evaluation of tolerance during intravenous administration of low dose of isosorbide dinitrate in the treatment of unstable angina. *Arch Mal Coeur Vaiss* 1992;85:59–65.
50. Thadani U. Nitrate intolerance: how to prevent it or minimize its effect. *Postgrad Med J* 1992;91:307–318.
51. DeMots H, Glasser SP, and the Transderm Nitro Study Group. Intermittent nitroglycerin therapy in the treatment of stable angina. *J Am Coll Cardiol* 1989;13:786–793.
52. Habbab M, Haft J. Heparin resistance induced by intravenous nitroglycerin: A word of caution when both drugs are used concomitantly. *Arch Intern Med* 1987;147:856.
53. Blasini R, Froer KL, Blumel KL, Rudolf W. Wirkungver lust von isosorbid dinitrat bei langzeibehandlung der chronischen herzinsuffizienz. *Herz* 1982;7:250.
54. Yusuf S, Sleight P, Rossi P, et al. Reduction in infarct size, arrhythmias and chest pain by early intravenous beta blockers in suspected acute myocardial infarction. *Circulation* 1983;67(Suppl I):I32–I41.
55. Quyyumi AA, Efthimiou J, Quyyumi A, et al. Nocturnal angina: Precipitating factors in patients with coronary artery disease and those with variant angina. *Br Heart J* 1986;56:346–352.
56. Grambow DW, Topol EJ. Effect of maximal medical therapy on refractoriness of unstable angina pectoris. *Am J Cardiol* 1992;70:577–581.
57. Thadani U. Role of nitrates in angina pectoris. *Am J Cardiol* 1992;70(Suppl):43B–53B.

Nitrates in Myocardial Infarction

Bodh I. Jugdutt

Summary. **Until two decades ago nitroglycerin was contraindicated in acute myocardial infarction (MI). Studies in the canine model demonstrated that low-dose intravenous (IV) infusion, carefully titrated to decrease mean blood pressure by 10% but not below 80 mmHg, during early stages of acute MI produced marked reduction of left ventricular (LV) preload, improvement in regional perfusion, and limitation of infarct size and remodeling. However, more IV nitroglycerin to decrease blood pressure further resulted in a paradoxical J-curve effect, with hypoperfusion and increased infarct size. Clinical studies have confirmed that low-dose IV nitroglycerin infusion for the first 48 hours after acute MI is safe, not only for improving performance in LV failure, but also for limiting ischemic injury, infarct size, remodeling, and infarct-related complications, including deaths in-hospital and up to 1 year. Recent studies suggest that more prolonged therapy with nitrates spanning the healing phase of acute anterior Q-wave MI can further limit LV remodeling and preserve function. Results of the recently completed ISIS-4 and GISSI-3 megatrials suggest, however, that long-term nitrate in patients with suspected MI in the 1990s does not improve survival significantly.**

Key Words. **intravenous nitroglycerin, low-dose infusion, infarct size, remodeling, mortality**

More than a century of learning and experience with the use of nitrates has led to the development of the low-dose intravenous (IV) nitroglycerin infusion regimen for acute myocardial infarction (MI) [1,2]. Before the 1970s nitrates were altogether avoided in acute MI because of the risk of severe hypotension and reflex tachycardia due to their powerful vasodilatory action [3,4]. In the early 1970s cardiologists began to use nitrates in congestive heart failure [5–7], including remote MI [8]. They placed emphasis on chronic reduction of LV impedance and afterload

[5–7], so that high doses and long-acting preparations were favored. Recognizing century-old warnings [3,4], other investigators began administering them cautiously for treating acute left ventricular (LV) failure of pulmonary edema complicating acute MI [9–11]. The search for a nitrate that has a short duration and a rapid onset and offset of action, can be accurately titrated, and can be promptly withdrawn in case of hypotension led to the use of IV nitroglycerin. While some physicians used IV nitroglycerin in high doses for controlling perioperative hypertension [12,13], others tried to develop a strategy for applying it safely to reduce ischemic injury and prevent LV failure in early stages of uncomplicated acute MI [14,15].

The concept of low-dose IV nitroglycerin infusion therapy for the management of acute MI therefore gradually developed between the late 1970s and early 1980s [1,2]. A large body of experimental and clinical evidence indicates that low-dose therapy is safe in acute MI and produces beneficial effects on multiple biological and clinical parameters with very little decrease in blood pressure and with little tolerance [1,2,16,17]. This paper reviews the evidence that such therapy is cardioprotective [1,3,16] and provides practical guidelines for its use in acute MI [18].

Mechanisms and Dosage Implications

The mechanisms of benefit from nitrates in acute MI are summarized in Table 1. The primary action of nitrates is direct vascular smooth muscle relaxation and dose-related vasodilation that leads to LV unloading and increased myocardial perfusion. In the dog model, very low doses of nitroglycerin in the range of 0.5 μg/kg/min [19] and equivalent to 35 μg/min for the 70 kg patient, or concentrations below 0.2 ng/ml [20] cause maximal peripheral venodilation that reduces venous return, LV size, wall tension, end-diastolic pressure and preload, and dilation of epicardial coronary arteries that increases perfusion. Tenfold higher doses, in the range of 5 μg/kg/min [19], or about 350 μg/min for the 70 kg patient, dilate peripheral arteries and decrease LV impedance, blood pressure, and afterload without further decreases in preload [20,21]. Fortyfold higher doses of 20 μg/kg/min [19], or 1400 μg/min for the 70 kg patient, dilate myocardial resistance vessels and further increase perfusion.

The marked decrease in preload seen with low doses, together with the associated decrease in LV filling pressure and wall tension, reduce sub-

Table 1. *Potential mechanisms for beneficial effects of nitrates in acute MI*

Improve hemodynamics
Decrease preload
Decrease afterload and impedance
Leftward and upward shift in LV function curve
Improve myocardial perfusion
Increase collateral blood flow
Decrease coronary artery spasm
Increase diameter of epicardial arteries and stenoses
Increase endothelium-derived relaxing factor (EDRF) activity
Increase prostacyclin activity
Decrease platelet adhesiveness
Increase coronary vein and lymphatic flow
Removal of noxious metabolites of ischemia
Decrease ischemic injury and infarct size
Improve cardiac geometry
Decrease chamber size, wall stress, and deformation
Decrease regional dilation and expansion
Decrease aneurysm formation
Decrease global dilation
Improve regional and global ventricular performance
Decrease infarct-related complications and improve survival

endocardial compression and improve perfusion [22]. Low doses also dilate atherosclerotic coronary arteries with eccentric stenoses, dilate large epicardial anastomotic collateral channels, and relieve coronary artery spasm [23]. In addition, low doses (6 μg/kg/min) dilate small intramural collateral vessels and increase collateral blood flow [24]. Furthermore, low doses decrease LV impedance and afterload in patients before a decrease in arm blood pressure can be detected by the sphygmomanometer [25]. The net result is improved myocardial energetics with decreased oxygen demand and increased supply.

The direct nitrate-induced vasodilation is endothelium independent and involves oxidation of sulfhydryl groups [26] and release of nitric oxide (NO) [27]. Depletion of the sulfhydryl groups is equated with tolerance [26], which is marked after chronic exposure to high doses of nitrates. It is now known that NO is identical to the endothelium-derived relaxing factor (EDRF) [28]. Since acute MI represents an EDRF- or NO-deficient state, nitrates might act by providing an exogenous source of EDRF or NO. Nitrates also produce vasodilation by an indirect, endothelium-dependent mechanism involving the release of prostacyclin (PGI_2) [29]. The combined effect of decreased EDRF, or NO, and PGI_2 is to promote platelet aggregation, release of vasoconstrictive substances such as thromboxane A_2 and serotonin, and coronary spasm in acute MI. Nitrates can potentially reverse or prevent these adverse effects by supplying EDRF or NO, and stimulating PGI_2 release [30]. In fact, nitrates given in low doses of 1.2 μg/kg/min, or about 70 μg/min, have been shown to produce significant antiplatelet and antithrombotic effects in acute MI [31].

In early stages of acute MI, nitrate-induced LV unloading and improved myocardial persufion can potentially decrease ischemic injury and infarct size [1]. As a consequence of decreased infarct size and the associated preservation of a normal subepicardial rim, less infarct expansion occurs and LV geometry is preserved [32]. In addition, acute LV unloading (by decreasing preload, afterload, chamber size, and wall stress) can potentially decrease mechanical distending forces acting on the infarct and noninfarct zones, limit infarct expansion and chamber dilation, and thereby preserve geometry [33]. During healing after acute MI, nitrate-induced LV unloading can also potentially prevent further remodeling [34]. Furthermore, the antithrombotic and antiplatelet effects can maintain patency of the infarct-related vessel after reperfusion therapies and prevent reocclusion, infarct extension, postinfarction angina, and/or recurrent infarction [30,31,35].

Cardioprotection and Myocardial Salvage

Controlled nitrate-induced vasodilation in early stages of acute MI reduces ischemic injury and decreases infarct size in experimental and clinical settings.

Experimental evidence

In a conscious dog model of inferior MI [24], infusion of low-dose IV nitroglycerin over the first 6 hours after left circumflex coronary artery ligation, to decrease mean arterial pressure by 10% but not below 90 mmHg, produced a 54% decrease in LV filling pressure or preload, a more than 50% increase in collateral blood flow, and a 51% decrease in infarct size measured pathologically at 2 days. The abolition of the 10% reduction in blood pressure by concomitant administration of methoxamine, an alpha-adrenergic agonist, resulted in similar beneficial effects on infarct size, preload, and collateral flow in that study [24]. These results suggested that the 10% decrease in blood pressure, or afterload, did not play a major role in the measured benefits. The clinical implication was that it is not always necessary, when giving IV nitroglycerin infusion in early stages of acute MI, to produce a decrease in blood pressure to ensure benefit.

In another conscious dog model of anterior MI [36], increasing the dose of IV nitroglycerin after left anterior descending coronary artery ligation to decrease blood pressure by more than 10% (range 17%–45%), down to levels below an average of 83 mmHg and as low as 58 mmHg, abolished the beneficial effects on collateral flow and infarct size, although the decrease in preload persisted. In fact, progressive nitroglycerin-induced afterload reduction produced a paradoxical J-curve effect on infarct size [2]. The dose had to be increased from 5 to 300 μg/min to achieve the graded reductions in blood pressure [36]. These results confirm that excessive vasodilator-induced hypoperfusion is potentially deleterious in early stages of acute MI. The clinical implications were that IV nitroglycerin infusion in early stages of acute MI was safe at a low dose titrated to a 10% decremental cutoff in blood pressure, and that more nitroglycerin is not necessarily better in acute MI and might increase infarct size.

In a subsequent study using the conscious dog model of anterior MI [37], early low-dose nitroglycerin limited not only infarct size but also infarct expansion. This preservation of LV geometry and limitation of postinfarct remodeling from a short pulse of nitrate therapy (during the first 6 hours after acute MI) underscores the importance of early therapy to prevent remodeling. In a similar model [34], a later pulse of nitroglycerin therapy between 2 and 7 days after anterior MI not only reduced infarct expansion and thinning over 6 weeks, but also improved LV function, improved mechanical strength, decreased distensibility, and increased the rupture threshold. Thus, nitrate therapy during and after acute MI has the potential of not only salvaging myocardium, but also geometry and function.

Clinical evidence

The randomized clinical trials of IV nitrates in acute MI [16,38–44] are summarized in Table 2. Seven of those used IV nitroglycerin infusion [16,38–41,43,44] and one European study used IV isosorbide dinitrate (ISDN) infusion [42]. None used thrombolytic therapy. All studies with IV nitroglycerin used low-dose infusion titrated to discrete hemodynamic endpoints (Table 2). However, titration endpoints differed, and regimens in trials between 1979 and 1984 [38–41,43,44] were non-uniform and often complicated. Invasive hemodynamic monitoring was often needed [41], but the blood pressure endpoint was reliably monitored by a Doppler device [13] or a

Table 2. *Titration regimens of IV nitrate in randomized trials in acute MI*

Reference	Number of patients	Interval from onset of AMI to therapy	Duration of therapy	Starting dose/ (range in μg/min)	Titration endpoint
Chiche et al. [38]	35 controls 35 NTG	<24 hr	5–7 days	15 μg/min	Reduce SBP by 20 mmHg
Bussmann et al. [39]	29 controls 29 NTG	<8 hr (mean 4.5) & >8 hr	48 hr	12.5 μg/min (37–60)	100 μg/min or reduce PCWP by 10–30% and MBP no more than 10%
Flaherty et al. [40]	42 controls 56 NTG	<12 hr	48 hr	5 μg/min (12–250)	Reduce MBP by 10%, keep SBP >90 mmHg, HR >50 bpm
Jugdutt et al. [41]	11 controls 11 NTG	<8 hr (mean 5.4)	48 hr	5 μg/min (5–120)	Reduce MBP by 10%, not <80 mmHg
Nelson et al. [42]	14 controls 14 ISDN	<14 hr	90 min (ISDN)	50 μg/kg/hr	200 μg/kg/hr or reduce MBP by 10 mmHg
Jaffe et al. [43]	42 controls 43 NTG	<12 hr	24 hr	10 μg/min (10–150)	200 μg/min to reduce SBP by 10% or to 95 mmHg, not <90 mmHg; HR not <50 bpm; HR increase not >20
Lis et al. [44]	76 controls 64 NTG	<12 hr	48 hr	10 μg/min (10–70)	Reduce SBP ≤30 in those with SBP >135 mmHg, ≤15 in those with SBP 120–134 mmHg, and ≤5 in those with SBP 95–119 mmHg (average dose 35 μg/min)
Jugdutt et al. [16]	156 controls 154 NTG	<12 hr (mean 5 hr)	48 hr	5 μg/min (4–192)	Decrease MBP by 10% in normotensive and 30% in hypertensive patients but not <80 mmHg

HR = heart rate; ISDN = isosorbide dinitrate; MBP = mean blood pressure; MI = myocardial infarction; NTG = nitroglycerin; PCWP = pulmonary capillary wedge pressure; SBP = systolic blood pressure.

sphygmomanometer [16]. The dose averaged between 29 and 90 µg/min (range, 4–250 µg/min) in the studies, which is much lower than those used in unstable angina pectoris and congestive heart failure.

The results of the randomized trials (Table 2) favored early therapy, preferably within 4 hours of onset of acute MI and continued for 24–48 hours. They confirmed findings from animal studies and showed that low-dose IV nitroglycerin therapy improved hemodynamics [16,38–41,43,44], decreased ischemic injury [16,39–41], reduced creatine kinase infarct size [16,39,41,43], reduced perfusion defect scores on thallium scintigraphy [40], improved regional LV systolic function with decreased endocardial surface area of mechanical dysfunction of asynergy [16,41], improved global LV systolic function [16,40,41], limited LV remodeling [16], decreased infarct-related complications [16,40], decreased morbidity [16,40,43], and decreased mortality in-hospital and at 1 year [16].

Low-Dose Regimen

In 1988 the results of a large randomized, placebo-controlled, single-blind clinical trial of IV nitroglycerin infusion in 310 patients were published [16] and established the safety and efficacy of a simple low-dose regimen in acute MI. The regimen (Table 3) is very similar to that recommended by the 1990 guidelines of the joint task force of the American College of Cardiology and the American Heart Association for the early management of patients with acute MI [18] and has become standard in several coronary care units in North America.

Briefly, the infusion is begun at a rate of 5 µg/min and titrated upward to achieve the decremental blood pressure target or a "ceiling" dose of 200 µg/min (when using standard IV sets and glass bottles). The dose is titrated to lower mean blood pressure by 10% in normotensive patients and by up to 30% in hypertensive patients (defined as those with blood pressure higher than 140/90 mmHg), but not below a mean blood pressure of 80 mmHg or a systolic pressure of 90 mmHg. Mean blood pressure is calculated simply as one third of the systolic pressure plus two thirds of the diastolic pressure. The blood pressure can be safely titrated and monitored by the nursing staff using a standard sphygmomanometer [1,16]. Blood pressure, heart rate, and clinical status are also monitored by the nursing staff throughout the infusion, which is maintained for at least 48 hours. The infusion is slowed down or temporarily stopped

Table 3. *Typical protocol for IV nitroglycerin infusion in acute MI*

1. Infusion via pump and standard or special IV set.
2. Start infusion with 5 μg/min.
3. Titrate upward by 5–10 μg/min every 5–10 min to the desired hemodynamic endpoint or a maximum of 200 μg/min.
4. Aim to lower mean BP by 10% in normotensives and 30% in hypertensives, but not <80 mmHg.
5. Monitor BP, heart rate, electrocardiogram, and clinical status.
6. Reassess if systolic BP <90 mmHg, mean BP <80 mmHg, diastolic BP increases >15 mmHg, or heart rate increases >20% or drops <50 bpm.
7. Maintain infusion for at least 24 hours, preferably 48 hours.
8. Stop the infusion if hypotension develops; resume infusion with 5 μg/min when safe.
9. To discontinue the infusion, titrate down gradually by 5–10 μg/min every 5–10 min.
10. The average dose range is 30–140 μg/min in acute MI.

BP = blood pressure; IV = intravenous; MI = myocardial infarction.

whenever the mean blood pressure falls below 80 mmHg, systolic blood pressure falls below 90 mmHg, or signs of excessive hypotension develop. After 48 hours the infusion is gradually tapered.

Using this regimen, Jugdutt and Warnica [16,17] found that the infusion could be safely continued for more than 12 hours in 89% of patients, more than 24 hours in 75%, and more than 48 hours in 27%. The average duration of infusion was 39 ± 25 hours, with the longest being 154 hours. The initial decrease in mean blood pressure was 10 ± 8% in normotensive and 19 ± 9% in hypertensive patients. In that study none of the patients developed hypotension requiring maneuvers to increase venous return or the administration of alpha-adrenergic agonists.

Is there a fixed or ideal dose?

There is no fixed or ideal dose of IV nitroglycerin infusion for acute MI. The initial infusion rate of 5 μg/min of the low-dose regimen is in sharp contrast with the fixed rate of 5 μg/kg/min or more used for congestive heart failure. Because of the dynamic progression of ischemic injury to necrosis and the unstable hemodynamics in acute MI, the dose should never be fixed. Because of differences in delivery systems, absorption into plastic and polyvinylchloride (PVC) tubing and bottles (but not glass), lack of a simple calculation to correct for absorption, and potential technical problems that can arise with improperly functioning infusion pumps, dose titration should always be used. Standard or commercially available non-PVC infusion sets, glass bottles, and a properly functioning infusion pump are recommended. Because of individual differences in dose response, sensitivity, and resistance, dose should always be titrated to the desired endpoint and subsequently adjusted.

Pain control and risk of hypotension

Doses more than 200 μg/min IV nitroglycerin in acute MI are associated with increased risk of hypotension [16–18]. High doses might precipitate more ischemic pain via a cycle of hypotension and ischemic injury. Beyond the decremental mean blood pressure cutoff of 80 mmHg, more nitroglycerin should not be given for "pain control" in acute MI. Titration to a hemodynamic endpoint (not "pain control") in acute MI has been shown to decrease ischemic injury [16,39,41,45,46]. In one study [46], a dose of 50 μg/min reduced ischemic injury in acute MI, but more or doubling of the dose (100 μg/min) produced tachycardia, hypotension, less decrease in ST-segment elevation, or an increase in ST-segment elevation.

In the large study by Jugdutt and Warnica [16], in which the average dose was only 45 μg/min, the blood pressure cutoff was successfully maintained in 91% of patients. However, in some patients (9% of 154 patients) whose mean blood pressure had inadvertently (without clinical symptoms of hypoperfusion) decreased below the 80 mmHg decremental endpoint in the first 12 hours of acute MI, the reduction in ischemic injury and infarct size was less than in those whose mean blood pressure remained above 80 mmHg. Despite these observations, multiple benefits were achieved [16]. Use of the low-dose regimen minimizes the risk of hypoperfusion and the paradoxical J-curve effect [2], which is especially critical in the first 12 hours of acute MI. The use of concomitant alpha-adrenergic agonists to counteract the decrease in blood pressure is not recommended because they can reverse the beneficial effects of IV nitroglycerin [45].

Increased sensitivity

Even when low-dose IV nitroglycerin infusion is used, there is a group that is at greater risk of hypotension, shows increased sensitivity to nitrates, and requires extra caution. These are patients with hypovolemia, right ventricular infarction, or inferior infarction with predominant right ventricular involvement, and the elderly.

Nitrate resistance

Marked initial refractoriness in lowering blood pressure, despite doses in excess of 200 μg/min, is commonly seen in studies of nitrates in congestive heart failure [1,2], suggesting primary receptor resistance and secondary mechanisms, such as catecholamine-induced increase in vasoconstrictor tone, activation of the renin-angiotensin system, hemodilution, impaired dilation due to vessel wall edema, and impaired hepatic metabolism of nitrate. Such refractoriness is also occasionally encountered in early uncomplicated acute MI, both in dogs [24,36] and humans [17,41], when high doses fail to decrease blood pressure, suggesting primary arterial and arteriolar resistance [17,41].

Who to treat?

In the large study of Jugdutt and Warnica [16], 154 of 310 consecutive patients who were admitted to the coronary care unit with acute MI within 12 hours (mean 5 ± 4 hours) of pain onset received nitroglycerin therapy. The patients had definite evidence of acute MI, mostly uncomplicated, their initial clinical Killip class being I in 47%, and II and III in 53%. Systolic blood pressure was more than 100 mmHg, heart rate below 120 but over 50 beats/min, and age below 75 years. Patients were excluded if they initially had cardiogenic shock, heart block, severe hypertension, or right ventricular infarction. Of the patients treated, 87% had Q-wave MI, 79% had a first MI, 45% had anterior MI, 55% had inferior MI, and 36% were hypertensive on admission.

Hemodynamic Effects

Nitroglycerin, as all nitrates, consistently decreases LV filling pressure, and the magnitude of the effect depends on the baseline level [9]. In acute MI, the beneficial effect on hemodynamics depends on several factors, some purely hemodynamic (e.g., baseline hemodynamic status, degree of hydration, and level of preload; dose and degree of afterload reduction) and others related to cardioprotection (effects on myocardial perfusion, ischemic injury, infarct size, LV geometry, and LV remodeling). The beneficial effect is achieved with a low dose. All studies in early MI (Table 2) showed marked decrease in LV filling pressure with less than 10% decrease in afterload [16,38–44]. However, the favorable effect on hemodynamics was partly due to the limitation of ischemic injury, infarct size, and remodeling [16,41].

Partial tolerance

In the study by Jugdutt and Warnica [16,17], mean blood pressure and the rate-pressure product remained lower for up to 10 hours with nitroglycerin

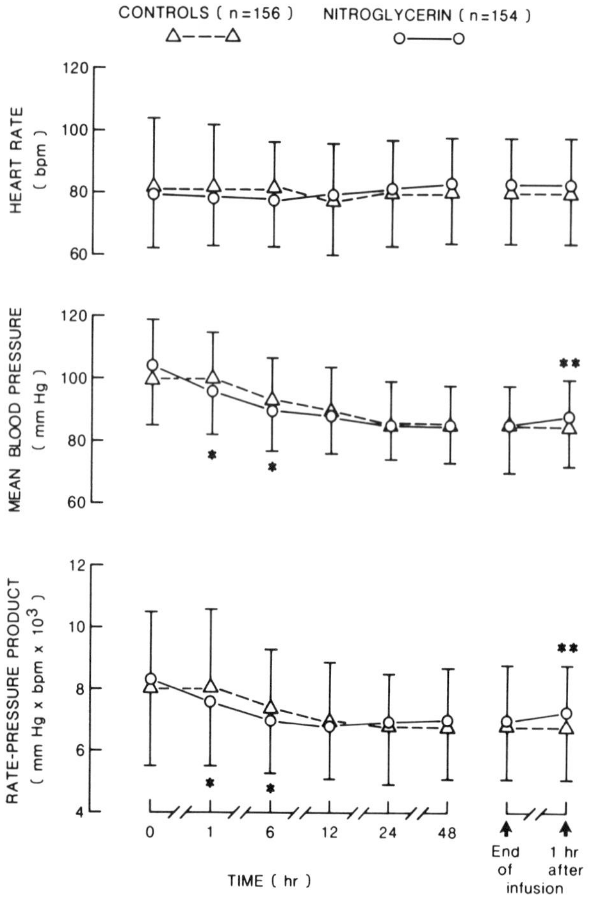

*Fig. 1. Hemodynamic changes in control and nitroglycerin groups during and after the infusions. *p ≤ 0.05, difference from control; **p ≤ 0.05, compared with value at the end of the infusion. (From Jugdutt and Warnica [16], with permission.)*

but did not differ from placebo between 12 and 48 hours (Figure 1). However, LV volumes and filling pressures remained lower in the nitroglycerin group, compared to placebo, for patients in whom these parameters were measured. Although the dose of nitroglycerin had to be increased by an average of 30 ± 39μg/min within 11 ± 9 hours after initial titration, individual analysis [17] revealed that this dose increase was needed for only 24% of the patients. More important, within 1 hour of stopping nitroglycerin, the effects on blood pressure, LV filling pressure, LV dimensions, and LV volumes were reversed (Figure 1). The hemodynamic tolerance found in 24% of the patients was, therefore, only partial [17].

Although this partial tolerance might have blunted the positive effect of nitroglycerin in infarct size, it did not blunt the beneficial effects on LV function and topography or complications [17]. It should be noted that beyond 12 hours of acute MI, the doses were slightly higher but the decremental mean blood pressure cutoff of 80 mmHg or the infusion rate of 200 μg/min were still not exceeded [16,17]. Whether a lower cutoff of mean blood pressure, for example, down to 70 mmHg, might be tolerated beyond the first 12 hours after acute MI has been suggested but not proven [12,17]. Greater percent reduction in blood pressure in the hypertensive patients was associated with similar benefits on LV hemodynamics and remodeling [16].

Ischemic Injury and Infarct Size

In the study by Jugdutt and Warnica [16,17], low-dose IV nitroglycerin infusion reduced ischemic injury, assessed by serial electrocardiographic ST-segment mapping, in both anterior and inferior MI, consistent with previous findings by Bussmann et al. [39], Flaherty et al. [40], and Jugdutt et al. [41]. More important, nitroglycerin therapy decreased cumulative creatine kinase infarct size in patients with anterior and inferior MI, and with both early and late therapy (Figure 2). Percent salvage was greater with early than late therapy, as also noted by Bussmann et al. [39]. The fact that salvage was observed in patients treated between 10 and 12 hours in this study [16], and in patients treated up to 14 hours by Flaherty et al. [40], suggests a wide therapeutic time window exists for obtaining benefit with low-dose IV nitroglycerin.

Left Ventricular Function

Low-dose IV nitroglycerin infusion in early acute MI improves LV function and shifts the relation between

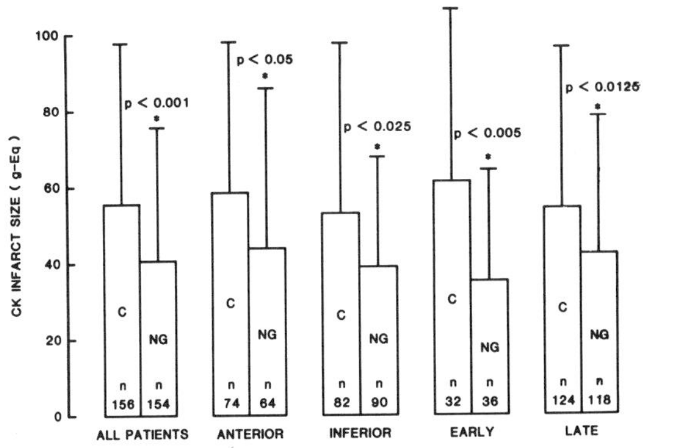

*Fig. 2. Creatine kinase (CK) infarct size in control (C) and nitroglycerin (NG) groups, anterior and inferior infarct subgroups, and early and late subgroups based on a 4-hour cutoff for therapy. *Significant difference comparing NG and C. (From Jugdutt and Warnica [16], with permission.)*

stroke volume and filling pressure leftward and upward [41]. What happens to stroke output depends on several factors, including the baseline preload, the shape of the Frank-Starling curve, and the degree of reduction in preload and afterload. In the randomized trials (Table 2), cardiac output either increased [16,39,41] or was maintained [38,40,42–44], but did not decrease. In the study by Jugdutt and Warnica [16], invasive hemodynamic monitoring was clinically indicated in 48 patients (29 placebo, 19 nitroglycerin). Nitroglycerin, but not placebo, produced a prompt decrease in pulmonary capillary wedge pressure from 20 ± 4 to 14 ± 3 mmHg and a late increase in cardiac index from 2.7 ± 0.6 to 3.1 ± 0.5 l/min/m^2 [16]. In the same study [16], serial two-dimensional echocardiograms in 128 patients (67 placebo, 61 nitroglycerin) showed a prompt decrease in the extent of LV akinesis and dyskinesis, and an increase in global ejection fraction (Figure 3). These benefits were evident at 10 days and 6 months, long after therapy was stopped [16]. The benefits were seen in both anterior and inferior MI, were greater in those treated before 4 hours of onset than after 4 hours of onset, and were less in those whose blood pressure dropped below 80 mmHg [16].

Left Ventricular Remodeling

In the study by Jugdutt and Warnica [16], low-dose IV nitroglycerin therapy decreased LV diastolic and systolic dimensions and volumes in anterior and inferior MI for up to a year. More important, this therapy limited infarct expansion and thinning (Figure 4). In fact, the placebo group showed a 31% increase in expansion index and a 17% decrease in thinning ratio from baseline to 10 days, but these parameters did not change in the nitroglycerin group (Figure 4).

Outcome and Survival

In the study by Jugdutt and Warnica [16], the positive effect of low-dose IV nitroglycerin therapy on LV remodeling was not only associated with smaller LV cavities with less regional deformation and better LV performance, but also with an 87% decrease in frequency of the clinical infarct expansion syndrome (from 15% to 2%), a 67% decrease in cardiogenic shock (from 15% to 5%), less congestive heart failure, and a 46% decrease of in-hospital mortality (from 26% to 14%) compared to placebo [16]. The functional Killip class score also improved with nitroglycerin and was 40% less than with placebo by 10 days (Figure 5). Furthermore, nitroglycerin produced

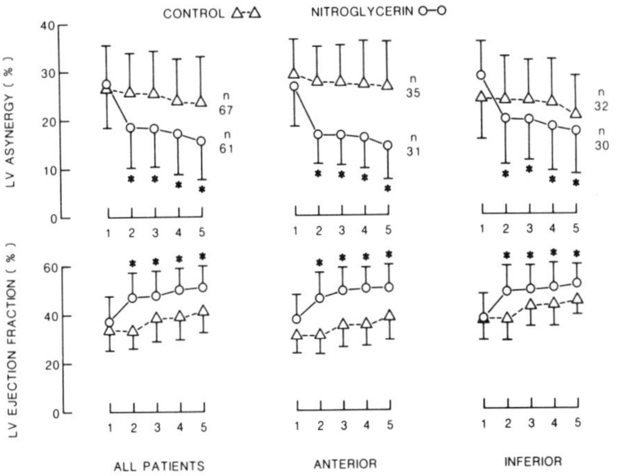

Fig. 3. *Changes in left ventricular (LV) asynergy and LV ejection fraction. *p < 0.001, significant difference of values at 6 (2), 24 (3), and 48 hours (4) and 10 days (5) from baseline value (1) by multiple measures analysis of variance. (From Jugdutt and Warnica [16], with permission.)*

a 77% reduction in frequency of LV thrombus on echocardiograms (from 22% to 5%) and a 50% reduction in infarct extension (from 22% to 11%). Nitroglycerin also decreased mortality at 3 months (from 28% to 16%) and 12 months (from 31% to 21%). This positive effect on mortality (Figure 6) was seen primarily in patients with anterior Q-wave MI who are at higher risk for remodeling [16].

Less infarct complications in the first 10 days after low-dose nitroglycerin in acute MI were also found in the trial by Flaherty et al. [40]. They showed, on retrospective analysis, that nitroglycerin in the early subgroup decreased the frequency of new congestive

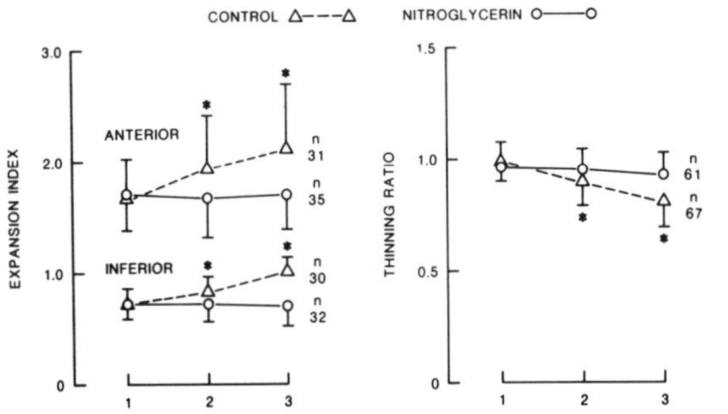

Fig. 4. *Changes in expansion index (left) and thinning ratio (right) for control and nitroglycerin groups. Expansion index is shown separately for the anterior and inferior infarct subgroups. *p ≤ 0.05, significance of difference comparing values at 2–3 days (2) and 10 days (3) with baseline. (From Jugdutt and Warnica [16], with permission.)*

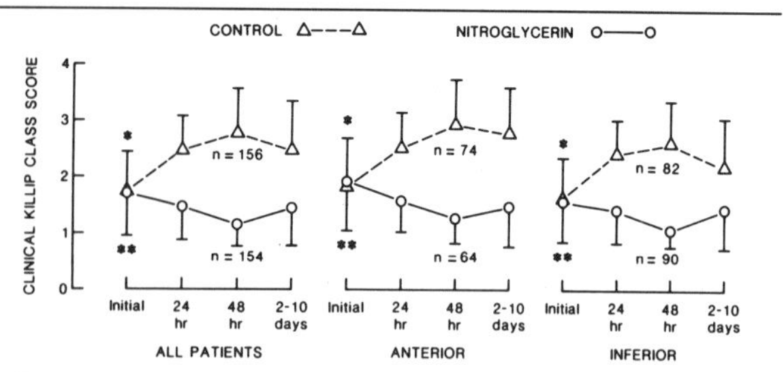

*Fig. 5. Changes in Killip class score in control and nitroglycerin groups. Values on admission (initial), at 24 and 48 hours, and the maximum between 2 and 10 days are shown. *p ≤ 0.05, significance of difference between baseline and subsequent values in all controls and control subgroups; **p ≤ 0.05, significance of difference between baseline and subsequent values in the nitroglycerin group and subgroups. (From Jugdutt and Warnica [16], with permission.)*

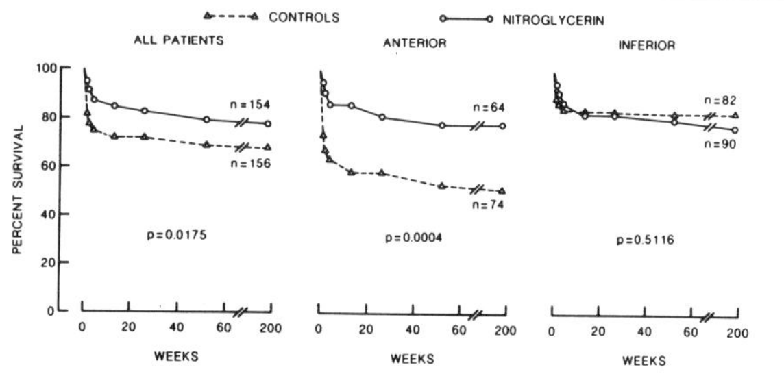

Fig. 6. Acturial survival curves for nitroglycerin and control groups. Data over the 43 months or 186 weeks of the long-term follow-up are shown. Left panel: All patients. Middle panel: Anterior subgroups. Right panel: Inferior subgroups. (From Jugdutt and Warnica [16], with permission.)

heart failure (from 30% to 6%), infarct extension (from 26% to 12%), death by 10 days (from 19% to 4%), and total in-hospital death (from 39% to 15%). The effect of low-dose IV nitroglycerin on mortality from all trials is summarized in Table 4. Because a large clinical trial of low-dose IV nitroglycerin therapy on survival in acute MI has not been performed, a meta-analysis of pooled data from North American and European trials was done by Yusuf et al. [47]. This showed a reduction in the odds of death of 49% with IV nitroglycerin and 35% for all nitrate trials (including three studies in which nitroprusside was also used).

Adjunctive Therapy

Very early reperfusion therapy is clearly effective in salvaging LV muscle, geometry, and function [32,33]. Late reperfusion, however, salvages geometry but

Table 4. *Mortality in major trials of IV nitroglycerin in acute MI*

Reference	Follow-up	Nitrate	Control	2p
Chiche et al. [38]	Hospital	3/50 (6%)	8/45 (18%)	NS
Bussmann et al. [39]	18 months	4/31 (13%)	12/29 (41%)	<0.02
Flaherty et al. [40]	3 months	11/56 (20%)	11/48 (23%)	NS
Jugdutt et al. [41]	Hospital	1/11 (9%)	2/11 (18%)	NS
Nelson et al. [42]	Hospital	0/14 (0%)	0/14 (0%)	NS
Jaffe et al. [43]	Hospital	4/57 (7%)	2/57 (4%)	NS
Lis et al. [44]	4 months	5/64 (8%)	10/76 (13%)	NS
Jugdutt et al. [16]	3 months	24/154 (16%)	44/156 (28%)	<0.01
Total	—	52/437 (12%)	89/436 (20%)	<0.001

might not salvage muscle and function [32,33]. Several preliminary reports indicate that low-dose IV nitroglycerin infusion can be safely combined with thrombolytic therapy in acute MI [48] and might widen the window for reperfusion therapy and accelerate recovery or recruitment of function after late reperfusion [48–52]. Furthermore, in high risk patients who have already developed marked infarct expansion very early in acute MI, aggressive LV unloading with the combination of low-dose IV nitroglycerin and intraarotic counterpulsation for 4–5 days can prevent further stretching of noninfarcted segments and LV dilation [53].

Sublingual Nitroglycerin

Sublingual nitroglyerin is used for ischemic pain upon initial presentation of the patient with acute MI to the emergency department. However, physicians should be alerted to the risk of producing excessive LV unloading, hypotension, and the paradoxical J-curve effect [1,2,16,36]. Topical and transdermal nitroglycerin might also have this effect in acute MI. Thus an excessive decrease in afterload is not cardioprotective and might impair perfusion; increase ischemic injury; worsen ischemic pain; increase infarct size; worsen hemodynamics; worsen regional and global LV function; increase infarct expansion; cause reflext tachycardia, which increases stretching of the infarct segment and lessens venodilation; and increase remodeling and LV dilation. Thus low-dose nitroglycerin infusion should be instituted early in acute MI, and if possible, in the emergency department itself.

Long-Acting Nitrates

It is a sound therapeutic principle to avoid long-acting pharmacologic agents in the early management of acute MI. Long-acting nitrates are therefore best avoided in that setting [18]. Although Nelson et al. [42] showed benefit with IV ISDN in acute MI, the drug was only given for 90 minutes and with careful hemodynamic monitoring, not allowing mean blood pressure to fall by more than 10 mmHg [42]. In unstable angina pectoris, both before and after acute MI, IV ISDN infusion has been very effective in "cooling off" the patient in Europe, although IV nitroglycerin is preferred in North America [31].

Prolonged Nitrate Therapy After Acute MI

The effect of LV unloading with prolonged nitrate therapy after acute MI has been evaluated in several clinical studies [49–52]. In the first randomized, placebo controlled, double blind study [49–51], 56 patients with a first anterior Q-wave acute MI were given low-dose IV nitroglycerin infusion for the first 48 hours followed by buccal nitroglycerin for 6 weeks (three times daily, with eccentric dosing at 8 AM, 1 PM, and 6 PM and an 8-hour nitrate-free interval to avoid vascular tolerance). The patients who were given long-term nitroglycerin therapy showed long-term preservation of LV geometry and improvement of LV function on echocardiography [49], persistently smaller LV volumes on radionuclide imaging [50], and prolongation of the benefits for up to 8.5 months following discontinuation of therapy [51].

Several other nitrate preparations (oral isosorbide dinitrate, oral isosorbide-5-mononitrate) and nitroglycerin patches are currently being evaluated for prolonged LV unloading and prevention of remodeling after acute MI. The angiotensin-converting-enzyme (ACE) inhibitor, captopril, has been shown to be highly effective in preventing postinfarct remodeling and increasing survival [54]. In a recently completed randomized, placebo-controlled study that was begun in 1988, the effects of prolonged therapy with captopril and nitroglycerin between 48 hours and 6 weeks after a first Q-wave acute MI were compared using a factorial design [52]. All patients received low-dose IV nitroglycerin infusion and thrombolytic infusion during the first 48 hours before randomization. The treatment regimens postrandomization consisted of captopril (6.25–12.5 mg three times daily orally), nitroglycerin (1–3 mg three times daily, buccal, eccentric dosing, 8 AM, 1 PM, and 6 PM, with 8 hour washout period), or matching placebo. The initial dose was titrated so that mean blood pressure did not fall by more than 10%. The preliminary results involved 80 patients and indicated that, compared to placebo, both nitroglycerin and captopril produced similar beneficial effects on LV volumes, LV asynergy, infarct expansion, infarct thinning, ejection fraction, aneurysm frequency, and morbidity parameters.

Megatrials of Nitrates Postinfarction

Previous trials of nitrates (Tables 2 and 4) were done in the prethrombolytic era and focused on high risk patients with MI. In contrast, several recent clinical trials [55–58] have explored the effect of prolonged nitrate therapy after suspected MI on mortality in the thrombolytic era. These recent studies, in a clearly different population with lower mortality, have suggested that chronic nitrate therapy might not benefit all patients in the 1990s.

A pilot substudy of the ISIS-4 trial by Pipilis et al. [55] showed that hemodynamic effects with isosorbide-5-mononitrate (20 mg three times daily in regular doses for 1 month) to patients with "suspected MI" were not sustained, whereas those with captopril (12.5 mg three times daily for 1 month) were sustained. Another pilot study of the ISIS-4 trial confirmed the safety of the treatment drugs, alone or in combination, in early MI [56]. The preliminary results of the ISIS-4 trial, which was designed to study the effect of very early IV magnesium followed by isosorbide mononitrate and/or captopril on mortality in patients with suspected MI [57], were recently presented at the American Heart Association meeting in November 1993 [58]. The ISIS-4 population that was tested for long-term nitrate differed from patients in earlier trials (Tables 2 and 4) in that low risk patients with inferior or no MI were included, the majority had already benefited from fibrinolytic and/or intravenous nitrate therapy and subsequently received other concurrent medications at the discretion of the responsible physician, and the 35-day mortality was low in the placebo group, varying from 6.92% to 7.33% [58]. The preliminary results [58] indicated that in such patients oral isosorbide-5-mononitrate (60 mg qid for 28 days) did not produce a statistically significant decrease in mortality (7.22 vs. 6.98%, 2p = NS), whereas oral captopril (50 mg bid for 28 days) produced a significant decrease (7.33% vs. 6.87%, 2p = 0.04), and IV magnesium sulphate over the first 24 hours produced a statistically nonsignificant increase in mortality (6.92% vs. 7.28%, 2p = NS). The trial was not designed to evaluate low dose IV nitroglycerin in acute MI or to study the effect on LV remodeling. The final results of the ISIS-4 trial [59], on 58,050 patients, confirmed that nitrates did not produce a clear decrease in mortality at 5-weeks or one year, although mortality in the first two days was somewhat less.

In the GISSI-3 study [60], a pilot trial compared the safety of IV nitroglycerin plus transdermal nitrate and oral lisinopril over the first 72 hours after suspected MI in 1521 patients. The potential for lowering blood pressure in the first 24 hours and producing hypoperfusion was shown to be greater with lisinopril than nitroglycerin [60], suggesting that "backing off" after hypotension might be a problem with lisinopril due to its longer half-life. The GISSI-3 trial assessed the effect of lisinopril (5 mg initially, then 10 mg qd), glyceryl trinitrate (GTN, intravenously for 24 hours, then 10 mg transdermal qd), or both, for 6 weeks postinfarction on survival and ventricular function. Preliminary results of the GISSI-3 study, briefly presented at the American Heart Association meeting of November 1993 [58], indicated that overall 42-day mortality was 6.7% and the combination of IV nitroglycerin in acute MI followed by nitrate patch for 42 days in patients with suspected MI in the thrombolytic era did not produce a significant decrease in mortality compared to the control group (6.5% vs. 6.9%, p = NS). Thus in nearly 80,000 patients with suspected acute MI, the combined preliminary results of the ISIS-4 and GISSI-3 trials suggested a small, statistically nonsignificant mortality reduction of 2.8 per thousand with prolonged nitrate therapy [58].

The final results of the GISSI-3 trial, on 19,394 patients, were published in May 1994 [61]. The results revealed that nitrates alone did not reduce mortality significantly but lisinopril reduced mortality by 11%. Importantly, the combination of lisinopril and nitrates produced a 17% reduction of total mortality (P = 0.02) and was not associated with excessive adverse events. The combination of lisinopril and nitrate was also highly effective in reducing the combined end-point of mortality and ventricular dysfunction in the elderly (21% reduction; P = 0.0004) and in women (21% reduction; P = 0.005). These results were found despite the fact that over half of the patients in the control groups (57 to 61%) received some kind of nitrate therapy. Furthermore, the nitrate group had a lower incidence of postinfarction angina (P = 0.03) and less cardiogenic shock (P = 0.009). The greater benefits of combined therapy in the GISSI-3 study suggest that the combination of ACE inhibitor and nitrates might have prevented tolerance.

The small added benefits found with nitrates alone in the ISIS-4 and GISSI-3 studies could be explained, in part, by development of early hypoperfusion, development of nitrate tolerance with chronic therapy, and heterogenous populations. However, the dominant factor appears to be that both populations were intensively exposed to optimum treatments and had low mortality. In contrast to the low mortality of about 7% for the control groups in these trials, the mortality in the control group of the early intravenous nitroglycerin study by Jugdutt and Warnica [16] was much higher, being 26% in hospital and 31% at one year. Although the patients in that study [16] were in a higher risk group, it is important to note that early intravenous nitroglycerin infusion decreased in-hospital mortality to 14% and one-year mortality to 21%, which are also much higher than the overall mortality found in the trials during the thrombolytic era. Clearly, the widespread application of thrombolysis, aspirin, beta-blockers and ACE-inhibitors have contributed to the decreased mortality after MI.

The overall results of the GISSI-3 study suggest that nitrates are beneficial in patients treated with thrombolytics and ACE-inhibitors, improving survival and decreasing ventricular dysfunction, especially in the elderly and in women. The use of intravenous nitroglycerin as adjunctive therapy during late reperfusion, to recruit ventricular function and further attenuate remodeling [48], requires further evaluation. Recent experimental evidence indicates that prolonged nitrate therapy, in various regimens during healing after non-reperfused MI, reduced left ventricular remodeling and dysfunction, and therapy over 6 weeks was more beneficial than therapy over the first 2 weeks [62]. Prolonged therapy with isosorbide-5-mononitrate, over 6 weeks after MI that were reperfused 2 hours after acute coronary occlusion in the dog, produced greater recovery of ventricular function and less remodeling than reperfusion alone [63]. Thus, judicious use of intravenous nitroglycerin in acute MI should be considered in patients who cannot receive reperfusion therapy and might still be beneficial in patients who receive reperfusion therapy.

Conclusions

In conclusion, LV unloading with low-dose IV nitroglycerin infusion can be safely applied in acute MI to preserve myocardium, geometry, and function, and might improve outcome and survival in high risk patients with anterior Q-wave MI. Nitrate tolerance is not a major problem with low-dose IV nitroglycerin therapy in early acute MI and can be avoided during long-term therapy by using appropriate formulations, eccentric dosing regimens, and an adequate nitrate-free interval. Early low-dose IV nitroglycerin and reperfusion, followed by prolonged therapy with ACE inhibitor or nitrate, or both, might prove to be an effective strategy for the management of high risk MI patients. However, results of the recent ISIS-4 and GISSI-3 megatrials in the thrombolytic era suggest that long-term nitrate in patients with suspected MI might not, but long-term ACE inhibitor might, improve survival in the 1990s.

Acknowledgments

This work was supported in part by grants from the Canadian Heart Foundation and the Medical Research Council, Ottawa, Ontario, Canada.

References

1. Jugdutt BI. Intravenous nitroglycerin infusion in acute myocardial infarction: Myocardial salvage. *Cardiovasc Rev Rep* 1984;5:1145–1163.
2. Jugdutt BI. Intravenous nitroglycerin unloading in acute myocardial infarction. *Am J Cardiol* 1991;68:52D–62D.
3. Prodger SH, Ayman D. Harmful effects of nitroglycerin: With special reference to coronary thrombosis. *Am J Med Sci* 1932;184:480–491.
4. Friedberg CK. Acute coronary occlusion and myocardial infarction. In: Friedberg CK, ed. *Diseases of the Heart, 3rd ed.* Philadelphia: WB Saunders, 1966:913–914.
5. Cohn JN. Vasodilator therapy for heart failure. The influence of impedance on left ventricular performance. *Circulation* 1973;48:5–8.
6. Taylor WR, Forrester JS, Magnusson P, et al. Hemodynamic effects of nitroglycerin ointment in congestive heart failure. *Am J Cardiol* 1976;38:469–473.
7. Chatterjee K, Parmley WW. The role of vasodilator therapy in heart failure. *Prog Cardiovasc Dis* 1977;19:301–325.
8. Gold HK, Leinbach RC, Sanders CA. Use of sublingual nitroglycerin in congestive heart failure following acute myocardial infarction. *Circulation* 1972;46:839–845.
9. Flaherty JT, Reid PR, Kelly DT, et al. Intravenous nitroglycerin in acute myocardial infarction. *Circulation* 1975;51:132–139.
10. Armstrong PW, Walker DC, Burton JR, et al. Vasodilator therapy in acute myocardial infarction: A comparison of sodium nitroprusside and nitroglycerin. *Circulation* 1975;52: 1118–1122.
11. Flaherty JT. Intravenous nitroglycerin in acute myocardial infarction. *Cardiovasc Rev Rep* 1990;11:46–52.
12. Káplan JA, Dunbar RW, Jones EL. Nitroglycerin infusion during coronary artery surgery. *Anesthesia* 1976;45:14–21.
13. Fahmy NR. Nitroglycerin as a hypotensive drug during general anesthesia. *Anesthesia* 1978;49:17–20.
14. Epstein SE, Kent KM, Goldstein RE, et al. Reduction of ischemic injury by nitroglycerin during acute myocardial infarction. *N Engl J Med* 1975;292:29–35.
15. Flaherty JT, Come PC, Baird MG, et al. Effects of intravenous nitroglycerin on left ventricular function and ST segment changes in acute myocardial infarction. *Br Heart J* 1976;38:612–621.
16. Jugdutt BI, Warnica JW. Intravenous nitroglycerin therapy to limit myocardial infarct size, expansion and complications. Effect of timing, dosage and infarct location. *Circulation* 1988;78:906–919.
17. Jugdutt BI, Warnica JW. Tolerance with low dose intravenous nitroglycerin therapy in acute myocardial infarction. *Am J Cardiol* 1989;64:581–587.
18. Gunnar RM, Fisch C. Special Report: ACC/AHA guidelines for the early management of patients with acute myocardial infarction. *Circulation* 1990;82:664–707.
19. Bassenge E, Stewart DJ. Effects of nitrates in various vascular sections and regions. *Z Kardiol* 1986;75(Suppl 3):1–7.
20. Imhof PR, Ott B, Frankhauser P, et al. Difference in nitroglycerin dose-response in the venous and arterial beds. *Eur J Clin Pharmacol* 1980;18:455–460.
21. Ferrer MI, Bradley SE, Wheeler HO, et al. Some effects of nitroglycerin upon the splanchnic, pulmonary and systemic circulations. *Circulation* 1966;33:357–373.
22. Dunn RB, Griggs DM Jr. Ventricular filling pressure as a determination of coronary blood flow during ischemia. *Am J Physiol* 1983;244:H429–H436.

23. Brown BG, Bolson E, Petersen RB, et al. The mechanisms of nitroglycerin action: Stenosis vasodilation as a major component of the drug response. *Circulation* 1981;64:1089–1097.
24. Jugdutt BI, Becker LC, Hutchins GM, et al. Effect of intravenous nitroglycerin on collateral blood flow and infarct size in the conscious dog. *Circulation* 1981;63:17–28.
25. Kelly RP, Gibbs HH, O'Rourke MF, et al. Nitroglycerin has more favorable effects on left ventricular afterload than apparent from measurement of pressure in a peripheral artery. *Eur Heart J* 1990;11:138–144.
26. Needleman P, Johnson EM Jr. Mechanism of tolerance development to organic nitrates. *J Pharmacol Exp Ther* 1973; 184:709–715.
27. Ignarro LJ, Lippton H, Edwards JC, et al. Mechanism of vascular smooth muscle relaxation by organic nitrates, nitroprusside and nitric oxide: Evidence for the involvement of S-nitrosothiols as active intermediates. *J Pharmacol Exp Ther* 1981;218:739–749.
28. Palmer RMJ, Ferrige AG, Moncada S. Nitric oxide release accounts for the biological activity of endothelium-derived relaxing factor. *Nature* 1987;327:524–526.
29. Trimarco B, Cuoculo A, Van Dorne D, et al. Late phase of nitroglycerin-induced coronary vasodilatation blunted by inhibition of prostaglandin synsthesis. *Circulation* 1985;71:840–848.
30. Lam JYT, Chesbro JH, Fuster V. Platelets, vasoconstriction and nitroglycerin during arterial wall injury. *Circulation* 1988;78:712–716.
31. Diodati J, Theroux P, Latour JG, et al. Effects of nitroglycerin at therapeutic doses on platelet aggregation in unstable angina pectoris and acute myocardial infarction. *Am J Cardiol* 1990;66:683–688.
32. Michorowski BL, Senaratne MPJ, Jugdutt BI. Myocardial infarct expansion. *Cardiovasc Rev Rep* 1987;8:42–48.
33. Michorowski BL, Senaratne MPJ, Jugdutt BI. Deterring myocardial infarct expansion. *Cardiovasc Rev Rep* 1987;8:55–62.
34. Jugdutt BI. Effect of nitroglycerin and ibuprofen on left ventricular topography and rupture threshold during healing after myocardial infarction in the dog. *Can J Physiol Pharmacol* 1988;66:385–395.
35. Becker RC, Gore JM, Alpert JS. Postinfarction unstable angina. Pathophysiologic basis for current treatment modalities. *Cardiology* 1989;76:144–157.
36. Jugdutt BI. Myocardial salvage by intravenous nitroglycerin in conscious dogs: Loss of beneficial effect with marked nitroglycerin-induced hypotension. *Circulation* 1983;68:673–684.
37. Jugdutt BI. Delayed effects of early infarct-limiting therapies on healing after myocardial infarction. *Circulation* 1985; 72:907–914.
38. Chiche P, Baligadoo SJ, Derrida JP. A randomized trial of prolonged nitroglycerin infusion in acute myocardial infarction (abstr). *Circulation* 1979;59, 60(Suppl II):II165.
39. Bussmann WD, Passek D, Seidel W, et al. Reduction of CK and CK-MB indexes of infarct size by intravenous nitroglycerin. *Circulation* 1981;63:615–622.
40. Flaherty JT, Becker LC, Bulkley BH, et al. A randomized prospective trial of intravenous nitroglycerin in patients with acute myocardial infarction. *Circulation* 1983;68:576–588.
41. Jugdutt BI, Sussex BA, Warnica JW, et al. Persistent reduction in left ventricular asynergy in patients with acute myocardial infarction by intravenous infusion of nitroglycerin. *Circulation* 1983;68:1264–1273.

42. Nelson GIC, Silke B, Ahuya RC, et al. Haemodynamic advantages of isosorbide dinitrate over frusemide in acute heart-failure following myocardial infarction. *Lancet* 1983; i:730–733.
43. Jaffe AS, Geltman EM, Tiefenbrunn AJ, et al. Reduction of infarct size in patients with inferior infarction with intravenous glyceryl trinitrate. A randomized study. *Br Heart J* 1983; 49:459–460.
44. Lis Y, Bennett D, Lambert G, et al. A preliminary double-blind study of IV nitroglycerin in acute myocardial infarction. *Intensive Care Med* 1984;10:179–184.
45. Come PC, Flaherty JT, Weisfeldt M, et al. Reversal of the beneficial effects of intravenous nitroglycerin in patients with acute myocardial infarction by phenylephrine. *N Engl J Med* 1975;293:1003–1007.
46. Bussmann WD, Schofer H, Kurita H, et al. Nitroglycerin in acute myocardial infarction: X. Effect of small and large doses of I.V. nitroglycerin on ST-segment deviation: Experimental and clinical results. *Clin Cardiol* 1979;2:106–112.
47. Yusuf S, Collins R, MacMahon S, et al. Effect of intravenous nitrates on mortality in acute myocardial infarction: An overview of the randomized trials. *Lancet* 1988;1:1088–1092.
48. Tymchak WJ, Michorowski BL, Burton JR, et al. Preservation of left ventricular function and topography with combined reperfusion and intravenous nitroglycerin in acute myocardial infarction (abstr). *J Am Coll Cardiol* 1988;11:90a.
49. Michorowski BL, Tymchak WJ, Jugdutt BI. Improved left ventricular function and topography by prolonged nitroglycerin therapy after acute myocardial infarction (abstr). *Circulation* 1987;76(Suppl IV):IV-128.
50. Humen DP, McCormick L, Jugdutt BI. Chronic reduction in left ventricular volumes at rest and exercise in patients treated with nitroglycerin following anterior MI (abstr). *J Am Coll Cardiol* 1989;13:25a.
51. Jugdutt BI, Neiman JC, Michorowski BL, et al. Persistent improvement in left ventricular geometry and function by prolonged nitroglycerin therapy after anterior transmural acute myocardial infarction (abstr). *J Am Coll Cardiol* 1990; 15:214a.
52. Jugdutt BI, Tymchak W, Humen D, et al. Prolonged nitroglycerin versus captopril therapy on remodeling after transmural myocardial infarction (abstr). *Circulation* 1990;82(Suppl III):III-442.
53. Flaherty JT, Becker LC, Weiss JL, et al. Results of a randomized prospective trial of intraaortic balloon counterpulsation and intravenous nitroglycerin in patients with acute myocardial infarction. *J Am Coll Cardiol* 1985;56:434–446.
54. Pfeffer MA, Braunwald E, Moye LA, et al. Effect of captopril on mortality and morbidity in patients with left ventricular dysfunction after myocardial infarction. *N Engl J Med* 1992;327:669–677.
55. Pipilis A, Flather M, Collins R, et al. Hemodynamic effects of captopril and isosorbide mononitrate started early in acute myocardial infarction: A randomized placebo-controlled study. *J Am Coll Cardiol* 1993;21:73–79.
56. Flather M, Pipilis A, Collins R, et al. ISIS-4 (Fourth International Study of Infarct Survival) Pilot Study Investigators. Randomized controlled trial of oral captopril, of oral isosorbide mononitrate and of intravenous magnesium sulphate started early in acute myocardial infarction: safety and haemodynamic effects. *Eur Heart J* 1994;15:608–619.
57. ISIS-4 collaborative group. Fourth International Study of Infarct Survival: protocol for a large simple study of the effects

of oral mononitrate, of oral captopril, and of intravenous magnesium. *Am J Cardiol* 1991;68:87D–100D.

58. ISIS collaborative group. ISIS-4: Randomized study of oral isosorbide mononitrate in over 50,000 patients with suspected acute myocardial infarction (Abstr). *Circulation* 1993;88:I-394.
59. ISIS-4: A randomized factorial trial assessing early oral captopril, oral mononitrate, and intravenous magnesium sulphate in 58,050 patients with suspected acute myocardial infarction. *Lancet* 1995;345:669–685.
60. GISSI-3 Investigators. GISSI-3 Pilot Study: Safety of lisinopril and nitroglycerin in the first 72 hours after acute myocardial infarction (Abstr). *Circulation* 1991;84(Suppl II):II-366.
61. Gruppo Italiano per lo Studio della Sopravvivenza nell' Infarcto Miocardico. GISSI-3: effects of lisinopril and transdermal glyceryl trinitrate singly and together on 6-week mortality and ventricular function after acute myocardial infarction. *Lancet* 1994;343:1115–1122.
62. Jugdutt BI, Khan MI. Effect of prolonged nitrate therapy on left ventricular remodeling after canine acute myocardial infarction. *Circulation* 1994;89:2297–2307.
63. Khan MI, Jugdutt BI. Functional impact of isosorbide-5-mononitrate therapy after late reperfusion in canine anterior myocardial infarction (Abstr). *Clin Res* 1993;41:233A.

Nitrates in Silent Ischemia

Henry Purcell, David Mulcahy, and Kim Fox

Summary. **In recent years it has become clear that episodes of transient myocardial ischemia commonly occur in patients with coronary artery disease in the absence of chest pain or angina equivalent. These episodes of "silent myocardial ischemia" are particularly well documented during continuous ambulatory electrocardiographic monitoring in daily life. Evidence suggests that these episodes represent true ischemia, and appear to be a marker of unfavorable outcome. While the pathophysiology is not completely understood, it appears as though the mechanisms of angina and silent ischemia are the same. Both forms of ischemia respond to conventional antianginal medication. While long-acting nitrates are effective in reducing or preventing myocardial ischemia, because of their propensity to cause tolerance they should be used intermittently and in association with either betablockers or calcium antagonists. Nitrates are safe and comparatively inexpensive, and will continue to play an important role in the treatment and prevention of angina. However, in the light of current knowledge, there is no specific indication for the treatment of silent ischemia by nitrates.**

Key Words. **silent myocardial ischemia, nitrates, oral, transdermal, efficacy, tolerance, rebound, circadian**

*I*t is clear that "silent" myocardial ischemia is common during exercise testing and during daily activities in patients with stable coronary artery disease (CAD) and following acute myocardial infarction. Evidence is accumulating that suggests painless myocardial ischemia carries similar prognostic weight as painful ischemia in patients with documented coronary disease. Prior to assessing the role of organic nitrates in the treatment and prevention of silent myocardial ischemia, it is important to understand its mechanisms, how best to identify it, and to establish its clinical relevance.

Silent Myocardial Ischemia

During exercise stress testing using a modification of the Master's technique, over 40 years ago Wood et al. [1] reported the occurrence of typically ischemic changes of ST-segment depression without pain. In discussing their observation, they suggested that it might be possible to detect "latent angina pectoris" by means of effort testing, before attacks occurred spontaneously. In 1974 Stern and Tzivoni, using the newer technology of continuous ambulatory electrocardiographic ST-segment monitoring in patients with stable angina, noted that episodes of transient ST-segment depression were common during routine activities, and that many such episodes occurred without pain. Subsequently, Deanfield et al. [3], using ambulatory monitoring in 30 patients, demonstrated that approximately 75% of all episodes of ischemia occurred without pain. Also, assessing regional myocardial uptake of rubidium-82 using positron emission tomography (PET), they showed evidence of tomographic defects during episodes of both symptomatic and asymptomatic ischemia, confirming that these episodes represented true ischemia. Subsequent functional and hemodynamic studies [4] in patients with CAD have confirmed that silent myocardial ischemia represents "true" ischemia.

Mechanisms of Myocardial Ischemia

Myocardial ischemia is a dynamic process associated with increased myocardial oxygen demand and decreased coronary supply. These are not the sole determinants [5], as catecholamines, various mediators of endothelium-dependent constriction and relaxation, and a host of vasoactive peptides are clearly fundamental to the development of myocardial ischemia. The main determinant of myocardial oxygen supply is the severity of obstructive atherosclerotic plaque in the coronary arteries. Blood flow may be further reduced by increased vasomotor tone at the site of coronary stenoses. The major determinants of myocardial oxygen demand are heart rate, blood pressure, ventricular volume, and contractility. In patients with stable angina, myocardial oxygen demand may be increased by exertion during treadmill exercise testing, which identifies the majority of patients with myocardial ischemia; however, as shown by Deanfield [3] and others, asymptomatic ST-segment changes observed by ambulatory monitoring during daily life occur commonly without any excessive increase in heart rate. Indeed, silent myocardial ischemia can be readily precipitated by mental stress in patients with coronary artery disease, at lower heart rates than exercise-induced ischemia [6]. There is also considerable evidence to support the role for a primary decrease in myocardial oxygen supply. We know that many episodes of transient ischemia do not appear to be precipitated by increases in determinants of demand, such as surges in heart rate. This is shown in a recent study comparing the characteristics of painful and silent ischemia, using ambulatory ST-segment monitoring in 288 patients with documented CAD and stable angina, 150 of whom were off, while 138 were on, therapy [7]. During 12,436 hours of monitoring, 890 ischemic episodes were recorded, of which 652 (73%) were silent. As shown in Table 1, silent and painful ischemic episodes were similar in terms of heart rate at onset of ischemia, increase in heart rate prior to ischemia, duration of ischemia, and percentage of episodes not preceded by heart rate increase. The only discernible difference between silent episodes and angina was greater maximal depression of the ST-segment during painful ischemic episodes. If anginal episodes reflected more severe ischemia, we might have expected the duration of ischemia with angina to have been longer than that seen during silent ischemia. This therefore supports the notion that "silent ischemia is painful ischemia without the pain," and that all ischemia during daily life might be regarded as equally important, irrespective of symptoms. It seems reasonable to consider too that both an increase in myocardial oxygen demand and a decrease in myocardial blood supply play a role in the genesis of silent ischemia, and that the relative contribution from both depends on treatment status, activity, and the triggers of ischemia. Further evidence to support this comes from the observation that currently available antianginal agents can prevent or modify silent as well as symptomatic ischemic episodes [8].

We still do not know why pain is not present during episodes of silent ischemia. Several mechanisms have been proposed, including the possibility that there is an alteration in the patient's sensibility or perception of pain, as found in diabetic neuropathy, for example. It remains controversial whether endorphins play a role in this hyposensitivity to pain. There is also the possibility that silent ischemia may represent less severe ischemia than the symptomatic form, but evidence from studies described above do not support this.

How Common is Silent Ischemia?

Cohn suggests that the presence of pain in individuals with CAD is just the "tip of the iceberg" [9], and that silent ischemia is ubiquitous. In a study of 150

Table 1. *Characteristics of silent and painful ischemia during ambulatory monitoring in patients with coronary arterial disease*

	Mean HR at ischemia (bpm)	Mean ↑ HR preischemia (bpm)	Episodes without ↑ HR (%)	Mean duration of ischemia (min)	Mean ST depression (mm)
Silent episodes (n = 652)	92.0	7.7	30	21.7	1.49
Painful episodes (n = 238)	93.9 (p = NS)	9.0 (p = NS)	27 (p = NS)	21.9 (p = NS)	1.73 ($p < 0.01$)

From Mulcahy et al. [7], with permission.

CAD patients [10] who were not selected on the basis of exercise test findings, and who were investigated off therapy, 73% of all ischemic episodes were silent, and approximately 50% of patients had evidence of silent ischemia during 48 hours of ambulatory monitoring. But 50% of all episodes of silent ischemia occurred in 7% of the study group, implying that frequent silent episodes were common in only a small number of patients. This finding has been confirmed in other studies [11,12], and it appears that the frequency of silent ischemia is 33–50%, with frequent episodes occurring in approximately 4–10% of patients with stable angina, depending on whether or not they are taking routine antianginal medication.

Does Transient Myocardial Ischemia Affect Prognosis?

Myocardial ischemia is perceived as bad. Recurrent brief episodes of myocardial ischemia may cause myocardial necrosis [13]. Also, structural alterations of the myocardium have been shown to develop in the endocardial layers of the transiently ischemic myocardium with normal function at rest [14]. In clinical terms, however, the prevalence of exercise-induced silent ischemia in the "apparently healthy" population is low; in one series it was 2.5% in men and 3.4% in women [15]. Results from other observations suggest that patients with asymptomatic coronary disease who have a positive exericse test for ischemia may have lower mortality than those patients with CAD and angina [16]. Until further data are available from studies of silent ischemia in totally asymptomatic individuals, this interpretation remains unconfirmed. Overall mortality for patients with stable angina is relatively low, however [17], and it appears that in most patients silent ischemia carries the same prognostic significance as angina pectoris [8,18,19].

Circadian Aspects of Transient Ischemia

An important aspect to consider in any overview of silent myocardial ischemia is the well-defined peak in occurrence among CAD patients during the early morning hours, shortly after waking [20]. Invariably a second, early evening peak in ischemic activity is observed too. As illustrated in Figure 1, this circadian pattern has also been shown in acute myocardial infarction [21] and sudden cardiac death [22]. Frenetic clinical research activity has attempted to ablate or attenuate these peaks with antiischemic drugs, with a view to preventing cardiovascular

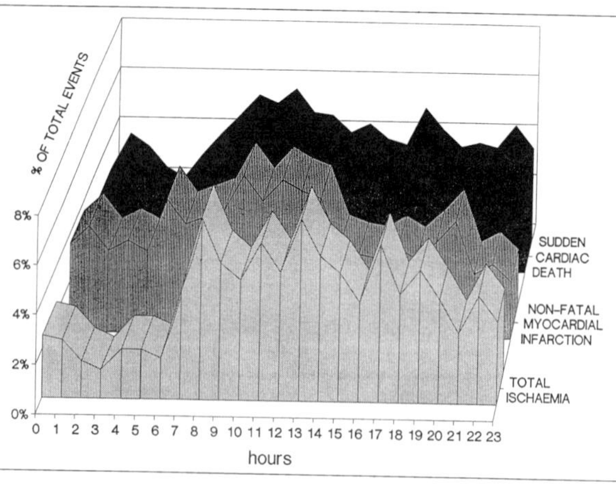

Fig. 1. *The circadian pattern of total ischemic episodes (598) in a study of 150 patients [20] when off therapy, of nonfatal myocardial infarction (n = 703) [21], and of sudden cardiac deaths [22]. Data are plotted as a percentage of total events during each hour of the day. (Reproduced with permission.)*

events in the early morning period [23]. While nitrates and calcium antagonists (used in routine clinical doses in the majority of controlled studies) have little effect, beta-blockers appear to blunt the morning peak in ischemia. In one study of 101 patients [24], a combination of metoprolol and nifedipine did abolish the morning and attenuate the evening peak in ischemic activity. At present we cannot say whether altering the circadian pattern of myocardial ischemia impacts on patient prognosis.

Therapeutic Implications

While several questions remain open about the prognostic significance of silent myocardial ischemia in stable angina, a case could be made for attempting to treat the total ischemic burden in patients with unstable angina or in the early phase following acute myocardial infarction, where transient ischemia appears to adversely affect outcome. But intervention among totally asymptomatic individuals is more difficult to justify [25,26]. One apparent cause for concern is the observation that silent ischemia occurs frequently in angina patients despite the use of conventional drugs to control systems [27]. Exercise-induced silent ischemia is also frequently seen early after successful coronary angioplasty [28]. Similarly, in a study of 36 patients studied before and after coronary artery bypass graft (CABG) surgery, Egstrup [29] found that, whereas surgery reduced the frequency of transient ischemia by 41% ($p <$

0.05), compared with medical therapy 12 patients actually developed more episodes of silent ischemia following their operation. A review of other surgical studies [30] suggests that more data are required before we can determine the significance of ST-segment shifts in patients following CABG surgery. However, accepting that the prognostic significance is uncertain, such findings imply that current strategies for prevention of myocardial ischemia, that is, antianginals and coronary revascularization, may be suboptimal.

Nitrates in Ischemic Heart Disease

Nitrates are a mainstay of therapy for the patient with ischemic heart disease. Most widely used sublingually in the form of a tablet or spray, nitrates may also be delivered orally in a sustained-release formulation, or via various long-acting delivery systems in the form of a buccal lozenge, as an ointment, in a transdermal patch, or intravenously. While nitrates are comparatively inexpensive, safe, and efficacious when used as short-acting preparations for relief of an acute anginal attack, the problem of tolerance (which will be reviewed in detail elsewhere) limits their use in angina prophylaxis. Nitrates may be used as monotherapy in patients who have minimal symptoms, but are invariably taken in combination with other antianginal preparations. One study [31] using ambulatory ECG monitoring in 72 CAD patients found that, when given as monotherapy, nitrates, beta-blockers, and calcium antagonists produced a 50% reduction in silent ischemic episodes, whereas when nitrates were combined with either of the two other agents there was a fourfold reduction in episodes compared to placebo. It may also be important to combine a nitrate with another class of agent to afford 24-hour anti-ischemic "cover," especially during the early morning period. Assuming, as the evidence provided suggests, that the pathophysiologic mechanisms of angina and silent myocardial ischemia are the same, nitrates offer an attractive means of preventing the occurrence of both forms of ischemia.

Nitrates in Silent Ischemia

While there is an extensive literature on nitrates in angina pectoris, there are comparatively few, large, well-controlled studies of nitrates in silent ischemia. In one of the earlier studies of silent ischemia during daily life [32], conducted using Holter monitoring in 20 CAD patients, 0.4 mg sublingual nitroglycerin administered at hourly intervals was shown to prevent or significantly reduce the frequency of ischemic episodes in a subgroup of five patients. While sublingual nitroglycerin provides rapid symptomatic relief of angina, its effects are short lived and it is therefore inappropriate for chronic use. Various oral preparations are now designed to have a longer duration of action. However, continuous therapy with either oral nitrates or nitroglycerin patches leads to rapid tolerance, which can be avoided with intermittent daily therapy [33]. Long-acting agents include isosorbide dinitrate (ISDN) and its metabolite isosorbide-5-mononitrate (IS-5-MN), which is almost 100% bioavailable after oral administration, and has the advantage of producing more predictable drug levels. Reiniger et al. [34] studied the effects of 120 mg ISDN in sustained-release formulation versus placebo, using 24-hour ambulatory ECG monitoring, in 25 CAD patients. During the placebo phase 277 episodes of ST-segment depression were detected, 81% of which were silent. The number of both symptomatic and silent ischemic episodes were reduced significantly to 119, that is, by −54% and −58%, respectively, as was the duration of ischemia (−29% and −51%, respectively). The daily distribution, in common with other studies, showed that 80% of ischemic episodes were incurred between 6 am and 6 pm, with two peaks in the morning and afternoon, and the drug did not influence this circadian pattern.

Dubiel et al. [35] performed a parallel, double-blind study, using slow-release ISDN, 120 mg daily, or placebo, in 32 patients with stable angina. Forty-eight-hour ambulatory ECG monitoring and exercise treadmill tests were done before and at 20 and 21 days of therapy. Analysis was performed on 15 patients in each group who completed the study. There was a significant decrease in the frequency of painful and silent ischemic episodes ($p < 0.001$) in patients receiving the nitrate; and also in the duration of painful and silent episodes. The magnitude of painful and silent episodes was also significantly reduced, as was time to ST-segment depression on exercise testing, in those receiving ISDN.

Von Arnim and Erath [36] have reported a 4-week, double-blind trial in CAD patients, comparing the effects of IS-5-MN 20 mg three times daily or 50 mg in a sustained-release tablet, or a nifedipine 20 mg sustained-release tablet three times daily. Both IS-5-MN and nifedipine reduced painful and painless ambulatory ischemic episodes to a similar extent, by about 60%.

Prognosis is adversely affected in patients with unstable angina [8] or acute myocardial infarction [18] who continue to experience episodes of silent ischemia. The effect of nitrates on transient ischemia have been assessed in both of these settings. Distante et al. [37] and Quyyumi [38] have shown that infusion of ISDN and oral mononitrate effectively reduce silent ischemia in both patients with vasospastic angina at rest and frequent nocturnal symptoms. Feng et al. [39] assessed the efficacy of IS-5-MN 20mg three times daily, versus placebo, on painful and silent ischemia, in 28 patients following myocardial infarction. The number of ischemic episodes was reduced from placebo, 538 to 64 (−88%); similarly, there were reductions in duration of ischemia (−94%) and total maximum ST-segment depression (−86%, $p < 0.01$) with IS-5-MN compared to placebo.

Transdermal Nitrates

The paucity of data relating to the effects of transdermal nitroglycerin on silent ischemia is alluded to in a recent overview by Pepine [40] assessing six studies, which are among those reviewed below. Shell [41] conducted an open study with transdermal nitrate in eight CAD patients, using ambulatory monitoring. A baseline tape was performed in the presence of beta-blockade and sublingual nitroglycerin. Transdermal nitroglycerin was then titrated in each patient, beginning with 5mg/24hr, and increasing the dose in increments of 5mg/24hr until ischemic events were abolished, when another Holter was performed. The mean dose used was 10.4mg/24hr. Eighty percent of events were silent. The total number of events fell from 5.3/24hr to 0.8/24hr ($p < 0.025$). Shell and Dobson [42] performed a double-blind, placebo-controlled, crossover comparison of nifedipine, 10mg three times daily, and transdermal nitroglycerin, 15mg, in 20 CAD patients. A 2-week placebo phase was followed by two 2-week treatment phases, with Holter monitoring and exercise testing at the end of each. The duration of ischemia was decreased by 57% (from 140min/24hr to 60min/24hr, $p = 0.0054$), and exercise tolerance time increased by 5.5% with transdermal nitroglycerin. With nifedipine, the duration of ischemia fell by 22% (from 175min/2hr to 137min/24hr, $p = 0.16$), while exercise time increased by 13%. The number of ischemic episodes was significantly reduced with both nitrate (−47%) and nifedipine (−39%), respectively. In addition to demonstrating the antiischemic efficacy of both agents, a further important finding from this study was that changes in exercise time do not necessarily predict changes in total ischemic time.

Schneeweiss and Marmor [43] demonstrated a significant reduction in ischemic episodes in a nonrandomized study of eight CAD patients, with the addition of transdermal nitrate 20–30 mg/hr, to a beta-blocker or calcium antagonist. Premature ventricular contractions were also reduced by 50%. Levine [44], in an open study of 26 patients, eight of whom completed the protocol that assessed a digital device for long-term ECG monitoring, showed that 10 mg of transdermal nitrate reduced the number of ischemic episodes by 59%, and the duration of ischemia by 60% ($p < 0.001$), with no apparent diminution of treatment effect over 2 weeks.

Three recent studies have compared the efficacy of continuous and intermittent transdermal nitrate therapy using ambulatory ECG monitoring in patients with CAD. Nabel et al. [45] studied 14 patients over 2 weeks, who demonstrated initial hemodynamic responsiveness to sublingual nitroglycerin, defined as a decrease in systolic blood pressure of >10 mmHg 10 minutes after sublingual nitrate administration. Only one patient remained on further antianginal therapy, a calcium antagonist, during the study. Patients then entered the titration phase, where transdermal nitrate patches were applied in 10 and 15 mg dose increments until a maximally tolerated dose of 30–60 mg/24 hr (mean 52 ± 5 mg) was achieved, and a decrease in systolic pressure of >10 mmHg was demonstrated. The individualized size and number of patches was then used throughout the double-blind, placebo-controlled, crossover phase. Two crossover phases were used, that is, continuous nitroglycerin (patches containing active drug worn for 24 hours) and intermittent therapy (12-hour active drug followed by 12-hour nitrate-free interval). A significant effect on the frequency of episodes (from 5.9 to 4.1/24 hr) was seen between placebo and treatment days of continuous therapy, but this was lost between 24 and 48 hours due to tolerance. Nonsignificant differences in frequency and duration of ischemia also occurred during intermittent therapy, which did not preserve the clinial efficacy of the preparation, however. This study has been criticized [47] on the grounds that all antiischemic therapy was withdrawn and some ischemic episodes may have been due to "rebound" ischemia; also, high doses were used for a relatively short period (48 hours).

Knuuti et al. [46] studied 85 patients, most of whom were taking a beta-blocker or calcium antagonist, in a randomized open trial over 12 weeks. Following a 3-week run-in period with continuous transdermal nitrate, 10 mg/24 hours, patients were randomized to either continuous or intermittent groups. In the intermittent group, patients removed their patch at night, the mean nitrate-free interval being 10 hours. Patients had Holter monitoring at randomization and after 2 and 12 weeks. There were 18 withdrawals, nine patients from each group. The number of ischemic episodes/48 hr with intermittent therapy was 3.1 after randomization, 1.8 at 2 weeks, and 2.0 at 12 weeks. With continuous therapy the figures were 3.8, 3.5, and 4.2, respectively. The differences were not statistically significant as some 30% of patients had no Holter ischemia. The number of ischemic episodes was shown to be statistically reduced, however, in 47 patients in the intermittent therapy group, ($p < 0.05$ at 12 weeks). The nighttime nitrate-free interval did not lead to increased ischemia (rebound). Based on these results, it appears that intermittent nitrate therapy has an additive antiischemic effect over concomitant medication and is superior to continuous therapy. It could not demonstrate whether continuous therapy was beneficial in preventing ischemia, however, as there was no placebo phase in the study. Also, it is not known whether tolerance developed during continuous therapy, as no comparable 48-hour Holter monitoring was performed during the screening phase of the study.

Fox et al. [47] studied 52 patients in a double-blind, multicenter, placebo-controlled study with two parallel limbs, who were randomized to receive either continuous 10 mg transdermal nitroglycerin ($n = 23$) or intermittent therapy with an 8-hour, nocturnal, drug-free interval and an individually titrated dose ($n = 29$) for 14 days. The intermittent regiment used a dose that was titrated to achieve a 10 mmHg fall in systolic blood pressure. If this was not achieved, another 10 mg/24 hr patch was applied, up to a maximum of four patches (40 mg/2 hr). Both treatments were compared with placebo in a crossover fashion. Treadmill exercise testing and Holter monitoring were performed after 14 days of treatment. Supine and standing systolic pressure fell by 7.5 mmHg and 9.0 mmHg, respectively with intermittent treatment ($p < 0.05$). Resting heart rate was unchanged, while mean heart rate at 1 mm ST depression rose by 11.9 beats/min, mean time to onset of angina increased by 59 seconds, and total exercise duration increased by 40 seconds (all $p < 0.05$). Treatment with continuous nitroglycerin did not affect these parameters. The frequency and circadian distribution of ischemic episodes was not reduced by either regimen. Intermittent treatment was not associated with an increase in nocturnal ischemia, and while this does not rule out rebound effects in isolated cases, it suggests that this phenomenon must be rare, if it occurs at all.

Rossetti et al. [48] recently reported a randomized, double-blind, placebo-controlled study of 12 patients with chronic stable angina, in which nitroglycerin patches (10 mg nitroglycerin) were worn either continuously or with overnight (8 hour) removal. Although the frequency of anginal attacks was reduced by both regimens, the effects on symptoms were not paralleled by any accompanying reduction in ischemic episodes during Holter monitoring. The authors concluded that the antiischemic effects of transdermal nitroglycerin, used as monotherapy, are probably too mild.

While transdermal nitroglycerin, given in sufficient doses, in the presence of beta-blockade can reduce the number, duration, and magnitude of ischemic episodes [49], this effect is not consistent when a transdermal nitrate is used as monotherapy. The optimal dose required for the elimination of all ischemic events is unknown. Approximately 2–4% of patients in studies with intermittent transdermal nitrate therapy have experienced rebound symptoms [50]. Although some studies have failed to demonstrate rebound ischemia following acute nitrate withdrawal, it is suggested that the coadministration of other antianginals, such as beta-blockers and calcium antagonists, may have masked it [51]. While it is widely accepted that a nitrate-free or nitrate-low period is required in order to avoid the problem of tolerance, we still do not know how long this should be. Newer nitrate preparations have been developed to address the problems of tolerance and rebound.

The availability of an oral, controlled-release formulation of IS-5-MN [52] that produces sufficiently high plasma concentrations during part of the day, followed by a low nitrate level, rather than a nitrate free interval, may have the potential to prevent both nitrate tolerance and rebound. Similarly, newer transdermal patches providing continuously increasing levels of glyceryl trinitrate may also limit the impact of tolerance and extend the efficacy period.

Conclusions

Silent myocardial ischemia is ubiquitous in patients with coronary artery disease. It may be attributable to both increases in myocardial oxygen demand and reduction in supply. It appears to carry an adverse prognosis, at least in the setting of unstable angina and in the early postinfarction period. Since nitrates decrease coronary artery tone and resistance, as well as myocardial oxygen requirements [53], they have the potential to relieve episodes of transient ischemia. Although evidence supporting the use of long-acting nitrates during daily ischemia is limited, a number of well-controlled clinical trials of systemic, sublingual, oral, and transdermal nitrates suggests that they can reduce episodes of silent ischemia. The only practical way to avoid development of tolerance, with loss of antiischemic efficacy, with nitrates is to adopt intermittent therapy [33]. However, intermittent therapy with oral or patch preparations leaves the patient unprotected at night and in the early morning hours.

Also, nitrates do not seem to affect the circadian pattern of myocardial ischemia. It is therefore advisable to use nitrates in combination with a long-acting beta-blocker or calcium antagonist [54]. While transient episodes of myocardial ischemia in stable angina may carry adverse prognostic significance, we do not know whether the treatment of silent, or indeed asymptomatic, episodes will positively affect prognosis. The recently reported Asymptomatic Cardiac Ischemia Pilot Study (ACIP) [55] is one of the current studies assessing the efficacy and safety of angina-directed medical therapy, angina plus ambulatory ECG-directed medical therapy, and revascularization. Patients were randomized to atenolol or diltiazem, in incremental doses, followed by two levels of isosorbide dinitrate. The primary endpoint was the elimination of ischemia on the 48-hour Holter ECG at 12 weeks. Findings from the pilot study suggest that a full-scale trial to see whether amelioration of silent ischemia improves survival is feasible. In the light of current knowledge there is no specific indication for the treatment of silent ischemia by nitrates.

References

1. Wood P, McGregor M, Magidson O, Whittaker W. The effort test in angina pectoris. *Br Heart J* 1950;12:363.
2. Stern S, Tzivoni D. Early detection of silent ischaemic heart disease by 24-hour electrocardiographic monitoring of active subjects. *Br Heart J* 1974;36:481–486.
3. Deanfield JE, Selwyn AP, Chierchia S, et al. Myocardial ischaemia during daily life in patients with stable angina: Its relation to symptoms and heart rate changes. *Lancet* 1983;1:753–758.
4. Levy RD, Shapiro LM, Wright C, Mockus L, Fox KM. Haemodynamic response to myocardial ischaemia during unrestricted activity, exercise testing, and atrial pacing assessed by ambulatory pulmonary artery pressure monitoring. *Br Heart J* 1986;56:12–18.
5. Collins P, Fox KM. Pathophysiology of angina. *Lancet* 1990;335:94–96.

6. Rozanski A, Bairey N, Krantz DS, et al. Mental stress and the induction of silent myocardial ischemia in patients with coronary artery disease. *N Engl J Med* 1988;318:1005–1012.
7. Mulcahy D, Keegan J, Fox KM. Characteristics of silent and painful ischaemia during ambulatory monitoring in patients with coronary arterial disease. *Int J Cardiol* 1990;28:377–379.
8. Pepine CJ. Is silent ischemia a treatable risk factor in patients with angina pectoris? *Circulation* 1990;82(Suppl II135–II142).
9. Cohn PF. *Silent Myocardial Ischemia and Infarction, 2nd ed*. New York: Marcel Dekker, 1989.
10. Mulcahy D, Keegan J, Crean P, et al. Silent myocardial ischaemia in chronic stable angina: A study of its frequency and characteristics in 150 patients. *Br Heart J* 1988;60:417–423.
11. Mulcahy D, Keegan J, Lindsay D, et al. Silent myocardial ischaemia in patients referred for coronary artery bypass surgery because of angina: A comparison with patients whose symptoms were well controlled on medical treatment. *Br Heart J* 1989;61:496–501.
12. Knight AA, Hollenberg M, London MJ, et al. Myocardial ischemia in patients awaiting coronary artery bypass grafting. *Am Heart J* 1989;117:1189–1195.
13. Mulcahy D, Phadke K, Eastick S, Walesby R, Davies M, Fox K. Recurrent transient episodes of myocardial ischemia: Effects at a cellular and clinical level. *Circulation* 1992;86(Suppl 1):1–116.
14. Hess OM, Schneider J, Nonogi H, et al. Myocardial structure in patients with exercise-induced ischemia. *Circulation* 1988;77:967–977.
15. Løchen M-L. The Tromso study; the prevalence of exercise-induced silent myocardial ischemia and relation to risk factors for coronary heart disease in an apparently healthy population. *Eur Heart J* 1992;13:728–731.
16. Epstein SE, Quyyumi AA, Bonow RO. Myocardial ischemia—silent or symptomatic. *N Engl J Med* 1988;318:1038–1043.
17. Mulcahy D, Parameshwar J, Holdright D, et al. Value of ambulatory ST segment monitoring in patients with chronic stable angina: Does measurement of the "total ischaemic burden" assist with management? *Br Heart J* 1992;67:47–52.
18. Detry JMR, Luwaert RJ, Melin JA. Prognostic significance of silent exertional myocardial ischaemia in symptomatic men without previous myocardial infarction. *Eur Heart J* 1992;13:183–187.
19. Tzivoni D, Gavish A, Zin D, et al. Prognostic significance of ischemic episodes in patients with previous myocardial infarction. *Am J Cardiol* 1988;62:661–664.
20. Mulcahy D, Keegan J, Cunningham D, et al. Circadian variation of total ischaemic burden and its alteration with antianginal agents. *Lancet* 1988;2:755–759.
21. Willich SN, Linderer T, Wegscheider K, Leizorovicz A, Alamercery I, Schroder R. Increased morning incidence of myocardial infarction in the ISAM study: Absence with prior beta-adrenergic blockade. *Circulation* 1989;80:853–858.
22. Muller JE, Ludmer PL, Willich SN, et al. Circadian variation in the frequency of sudden cardiac death. *Circulation* 1987;75:131–138.
23. Purcell H, Mulcahy D, Fox K. Circadian patterns of myocardial ischaemia and the effects of antianginal drugs. *Chronobiol Int* 1991;8:309–320.
24. Egstrup K. Attenuation of circadian variation by combined antianginal therapy with suppression of morning and evening

increases in transient myocardial ischemia. *Am Heart J* 1991;122:648–655.

25. Barsotti A. Silent myocardial ischaemia; to treat or not to treat? *Eur Heart J* 1991;12(Suppl G):8–15.
26. Mulcahy D, Fox K. Can we really justify the treatment of silent ischemia in 1992? No! *Cardiovasc Drugs Ther* 1992;6:125–129.
27. Deedwania PC, Carbajal EV. Prevalence and patterns of silent myocardial ischemia during daily life in stable angina patients receiving conventional antianginal drug therapy. *Am J Cardiol* 1990;65:1090–1096.
28. Kaul U, Dev V, Manchanda SC, Wasir HS. Silent myocardial ischemia after percutaneous transluminal coronary angioplasty and its prognostic significance. *Clin Cardiol* 1991;14:563–566.
29. Egstrup K. Asymptomatic myocardial ischemia as a predictor of cardiac events after coronary artery bypass grafting for stable angina pectoris. *Am J Cardiol* 1988;61:248–252.
30. Bertolet BD, Hill JA, Pepine CJ. Treatment strategies for daily life silent myocardial ischemia: A correlation with potential pathogenic mechanisms. *Proc Cardiovasc Dis* 1992;2:97–118.
31. Koehn DK, Glasser SP. The impact of antianginal drug therapy on asymptomatic myocardial ischemia. *J Clin Pharmacol* 1989;29:722–727.
32. Schang SJ, Pepine CJ. Transient asymptomatic ST segment depression during daily activity. *Am J Cardiol* 1977;39:396–402.
33. Thadani U. Role of nitrates in angina pectoris. *Am J Cardiol* 1992;780:43B–53B.
34. Reiniger G, Lehmann G, Beyerle A, Wolf H, Rudolf W. Treatment of symptomatic and asymptomatic myocardial ischemia during everyday activities with 120 mg isosorbide dinitrate in sustained release form [German]. *Herz* 1990;15:259–265.
35. Dubiel JP, Krzysztof W, Moczurad W, Bryniarski L. Efficacy of a single dose of slow-release isosorbide dinitrate in the treatment of silent or painful myocardial ischemia in stable angina pectoris. *Am J Cardiol* 1992;69:1156–1160.
36. Von Arnim T, Erath A. Nitrates and calcium antagonists for silent myocardial ischemia. *Am J Cardiol* 1988;61:15E–18E.
37. Distante A, Maseri A, Severi S, Biagini N, Chierchia S. Management of vasospastic angina at rest with continuous infusion of isosorbide dinitrate: A double crossover study in a coronary care unit. *Am J Cardiol* 1979;44:533–539.
38. Quyyumi AA, Crake T, Wright C, Mockus L, Fox KM. Medical treatment of patients with severe exertional and rest angina; double-blind comparison of beta-blocker, calcium antagonist and nitrate. *Br Heart J* 1987;57:505–511.
39. Feng J, Feng X-H, Schneeweiss A. Efficacy of isosorbide-5-mononitrate on painful and silent myocardial ischemia after myocardial infarction. *Am J Cardiol* 1990;65:32J–35J.
40. Pepine CJ. Daily life ischemia and nitrate therapy. *Am J Cardiol* 1992;70:54B–63B.
41. Shell WE. Mechanisms and therapy of silent myocardial ischemia and the effect of transdermal nitroglycerin. *Am J Cardiol* 1985;56:231–271.
42. Shell WE, Dobson D. Dissociative of exercise tolerance and total ischemic burden in chronic stable angina pectoris. *Am J Cardiol* 1990;66:42–48.
43. Schneeweiss A, Marmor A. Transdermal nitroglycerin patches for silent ischemia during antianginal treatment. *Am J Cardiol* 1988;61:36E–38E.
44. Levin RI. Quantitation of transient myocardial ischemia by

digital, ambulatory electrocardiography. *Am J Cardiol* 1988;61:13B–17B.

45. Nabel EG, Barry J, Rocco MB, Mead K, Selwyn AP. Effects of dosing intervals on the development of tolerance to high dose transdermal nitroglycerin. *Am J Cardiol* 1989;63:663–669.
46. Knuuti MJ, Wahl M, Wiklund I, et al. Acute and long-term effects on myocardial ischemia of intermittent and continuous transdermal nitrate therapy in stable angina. *Am J Cardiol* 1992;69:1525–1532.
47. Fox KM, Dargie HJ, Deanfield J, Maseri A, on behalf of the Transdermal Nitrate Investigators. Avoidance of tolerance and lack of rebound with intermittent dose titrated transdermal glyceryl trinitrate. *Br Heart J* 1991;66:151–155.
48. Rossetti E, Luca C, Bonetti F, Chierchia SL. Transdermal nitroglycerin reduces the frequency of animal attacks but fails to prevent silent ischemia. *J Am Coll Cardiol* 1993;21:337–342.
49. Shell WE, Kivowitz CF, Rubins SB, See J. Mechanisms and therapy of silent myocardial ischemia: The effect of transdermal nitroglycerin. *Am Heart J* 1986;112:222–229.
50. Cowan C. Clinical tolerance. In: Nitric Oxide—the Rational Basis for Nitrate Therapy. Wiliam Harvey Research Conferences (UK), Postgraduate Medical Education Series, Vol. 1, 1992:51–52.
51. Maxwell SRJ, Kendall MJ. An update on nitrate tolerance: Can it be avoided? *Postgrad Med J* 1992;68:857–866.
52. Olsson G, Allgen J. Is there an optimal prophylactic nitrate therapy? *Eur Heart J* 1991;12(Suppl A):21–23.
53. Chatterjee K. Role of nitrates in silent myocardial ischemia. *Am J Cardiol* 1987;60:18H–25H.
54. Rubboli A, Sangiorgio P, Camillo Pavesi P, et al. Efficacy of long-term administration of transdermal nitroglycerin in asymptomatic patients with effort-induced silent myocardial ischemia. *Cardiology* 1994;84:247–254.
55. Knatterud GL, Bourassa MG, Pepine CJ, et al. Effects of treatment strategies to suppress ischemia in patients with coronary artery disease: 12 week results of the Asymptomatic Cardiac Ischaemia Pilot (ACIP) Study. *J Am Coll Cardiol* 1994;24:11–20.

Nitrates in Congestive Heart Failure

Jocelyn Dupuis

Summary. **Nitrates are commonly used in the therapy of congestive heart failure (CHF). They exert beneficial hemodynamic effects by decreasing left ventricular filling pressure and systemic vascular resistance while modestly improving cardiac output. The improvement in left ventricular function caused by nitrates is the result of combined reduction in outflow resistance and mitral regurgitation, while decreased pericardial constraint and subendocardial ischemia may also contribute to the process. With continuous nitrate administration, complete arterial tolerance develops, while venous tolerance appears to be only partial. The major mechanism of tolerance is loss of vascular smooth muscle sensitivity to nitrates. An increase in total blood volume occurring during the first few hours of an acute administration may partly contribute to tolerance. The importance of reflex neurohumoral activation is controversial; although it may contribute to tolerance in CHF, its role does not appear to be major. Chronic continuous nitrate therapy in CHF improves submaximal and maximal exercise tolerance. In combination therapy with hydralazine, isosorbide dinitrate reduces mortality, although to a lesser extent than the angiotensin converting enzyme inhibitor enalapril. Intravenous or sublingual nitrates are first-line agents in the therapy of acute pulmonary edema. In severe CHF, refractory to standard medical therapy, a short course of intravenous nitroglycerin, with or without inotropic agents, can help break the vicious spiral of CHF. Because tolerance occurs without nitrate-free intervals and until an optimal schedule of administration is determined, it makes good sense to include a nightly nitrate-free interval when prescribing nitrates for CHF in order to maintain maximal benefit during the hours of activity.**

Key Words. **nitrates, congestive heart failure, nitrate tolerance, pulmonary edema, vascular volume**

Nitrates were the first vasodilators used for the therapy of congestive heart failure [1]. Because of excellent oral availability, innocuousness, and a hemodynamic profile similar to that of nitroprusside, nitrates were regarded as potential candidates in the vasodilator therapy of congestive heart failure [2–4]. Although numerous studies have assessed the efficacy of nitrates in congestive heart failure, their interpretation is rather complex, and no clear consensus has been achieved regarding many aspects of their clinical use. This difficulty arises from the wide range of doses and formulations used, the varying functional class of patients in one study versus another, and the lack of large, randomized placebo controlled trials [5]. In this paper we will review the acute and chronic hemodynamic effects of nitrates, their usefulness in various congestive heart failure states, and possible mechanisms of tolerance. Relevant data of the influence of nitrates on exercise capacity and mortality will also be reviewed, and recommendations based on our current knowledge will be made.

Acute Hemodynamic Effects

The acute use of nitrates can be defined as the first few hours of nitrate administration following a nitrate-free interval of at least 8 hours. There remains no controversy as to the beneficial acute hemodynamic alterations caused by nitrates in heart failure. With all modes of administration (oral, transdermal, sublingual, intravenous, buccal), the acute use of nitrates causes marked venodilatation and thus a substantial and significant decrease in pulmonary capillary wedge pressure as well as pulmonary artery and right atrial pressures [6–15]. Arterial dilatation, and thus the decrease in systemic arterial pressure and systemic vascular resistance, are characteristically less pronounced. Heart rate remains stable while cardiac output and derived indices usually increase mildly (Figure 1). The relative improvement in cardiac output appears to be directly proportional to the severity of heart failure (low cardiac index and high filling pressures) [12,16]. Of note is that these effects are dose dependent [17] and usually require larger doses of nitrates than those commonly used for angina therapy [15,18–20].

Numerous mechanisms have been postulated to explain the improvement in cardiac function caused by nitrates. The relative importance of each of these remains to be determined [21]. By virtue of their modest arteriolar dilatation, nitrates reduce afterload and thus can improve cardiac output. Additionally, nitrates can acutely reduce mitral regurgitant volume by as much as 50% with a proportional increase in forward output [22]. A greater degree of mitral regur-

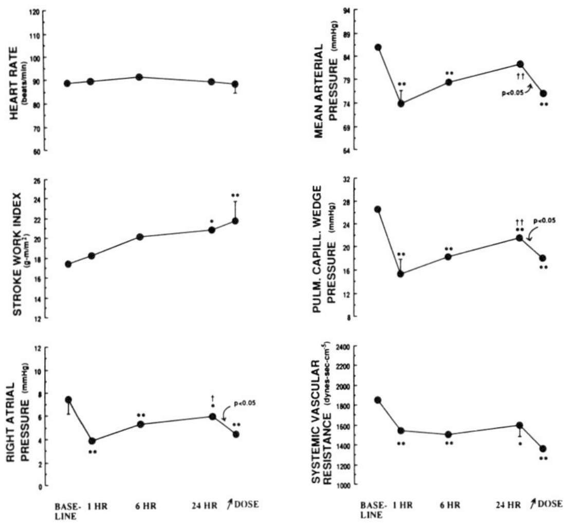

Fig. 1. *Hemodynamic effect of a continuous 24-hour infusion of nitroglycerin in 13 patients with class III and IV congestive heart failure. Values are shown at baseline and after 1, 6, and 24 hours of therapy with 1.5 µg/kg/min IV nitroglycerin. At the end of infusion the dose was increased to 7.5 µg/kg/min. Complete arterial tolerance developed, while there was only partial venous tolerance. *$p < 0.05$ and **$p < 0.01$ versus control; †$p < 0.05$ and ‡$p < 0.01$ versus 1 hour. (Reproduced in part from Dupuis et al. [7], with permission.)*

gitation and higher sensitivity to variations of afterload in severe heart failure have both been proposed as the mechanisms by which nitrates have a greater effect as heart failure progresses [23,24]. Nitrates also apparently improve left ventricular compliance on pressure-volume curves. This likely results from redistribution of blood volume from the heart and lungs to the periphery and the mesenteric bed [25,26], thus decreasing filling pressures and cavity volumes, which in turn reduces pericardial constraint. Finally, nitrates may improve myocardial contractility by reducing subendocardial ischemia and reducing myocardial oxygen requirements [27–29]. These additional benefits would be especially valuable in patients with associated coronary artery disease.

Chronic Hemodynamic Effects

After 6 hours of continuous nitrate administration, there is attenuation of the initial hemodynamic effects, resulting in a partial increase of all pressures (Figure 1). Tolerance develops between 12 and 24 hours of continuous administration [6,7]. All investigators agree that arterial tolerance to nitrates is complete in heart failure, while the degree of venous tolerance is still debated. In many trials, venous tolerance is partial with pulmonary capillary, pulmonary artery, and right atrial pressures still significantly lower than baseline, even after 24 hours of continuous administration [6,7,19,20,30–32]. This beneficial effect has been maintained for 72 hours with continuous intravenous nitroglycerin [6] and for up to 3 months with oral isosorbide dinitrate (40mg four times a day) [33,34,35]; and transdermal nitrates (5mg/24hr and 10mg/24hr given continuously) [32]. Unfortunately, the wide disparity in doses and forms of nitrates used and the small number of patients in many studies further complicates the interpretation and comparison of these studies. However, when scrutinized, most studies evaluating the effects of continuous nitrate administration show filling pressures that are chronically maintained at a lower level, although sometimes non-significantly [15,18,36]. It is the authors' opinion that nitrates, even when given continuously, maintain a sustained partial beneficial hemodynamic effect on venous pressures, and that this is likely responsible for the symptomatic improvement of congestive heart failure patients on chronic therapy [6]. Whether chronic intermittent therapy would be more effective than chronic continuous therapy remains to be determined. This issue will be discussed further in this review.

Tolerance

In patients with congestive heart failure, tolerance has been demonstrated with all forms of nitrates. It appears that in heart failure, venous tolerance is only partial while arterial tolerance is complete. Tolerance is maximal after 16–18 hours of continuous administration. The possible mechanisms of tolerance are multiple and their interplay complex. The possible mechanisms of tolerance to nitrates in congestive heart failure are shown in Figure 2. Decreased smooth muscle sensitivity to nitrates is the major mechanism of tolerance (end organ tolerance). The cause of this decreased sensitivity is unknown. One proposed mechanism has been smooth muscle sulfhydryl group depletion. However, although sulfhydryl group donors can transiently potentiate the effect of nitrates when given as a large single dose [13], it would appear that this is due to an extracellular reaction with nitrates [37], and chronic or continuous administration of sulfhydryl group donors has failed to prevent tolerance in congestive heart failure [7]. Presently, there is no indication that the addition of a

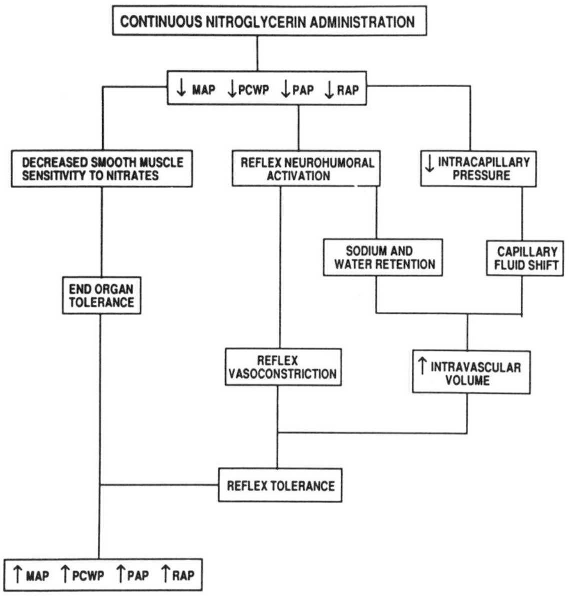

Fig. 2. *Possible mechanisms of nitrate tolerance in congestive heart failure. End-organ tolerance is the major mechanism of tolerance. The occurrence of reflex neurohumoral activation is debated. A nitrate-induced capillary fluid shift may partly contribute to tolerance in the first few hours of administration. MAP = mean arterial pressure; PCWP = pulmonary capillary wedge pressure; PAP = pulmonary artery pressure; RAP = right atrial pressure.*

sulfhydryl group donor to nitrates will be of value in the therapy of congestive heart failure.

A second potential mechanism of tolerance is an increase in intravascular volume. Nitrate administration causes a significant and rapid decrease in blood hematocrit that is maximal after 24 hours of administration (Figure 3). This rapid hemodilution appears to be the result of what we have termed a *capillary fluid shift* following Starling forces [7]. This capillary shift occurs within the first few hours of initiating nitrate administration and results in an estimated increase in intravascular volume of 14%. Dupuis et al. [7] measured a mean increase of 745ml in the total blood volume of heart failure patients, the majority of which occurred during the first 5 hours of nitrate administration. This capillary fluid shift may contribute to early tolerance and could help explain the rapid onset of action of nitrates in acute pulmonary edema: Although hypothetical, a shift of fluid from the alveolar space to the vasculature would rapidly relieve dyspnea. This nitrate associated hemodilution has now been reproduced by others in heart failure [38] and angina patients [39]. With longer continuous nitrate administration, fluid retention, as has been

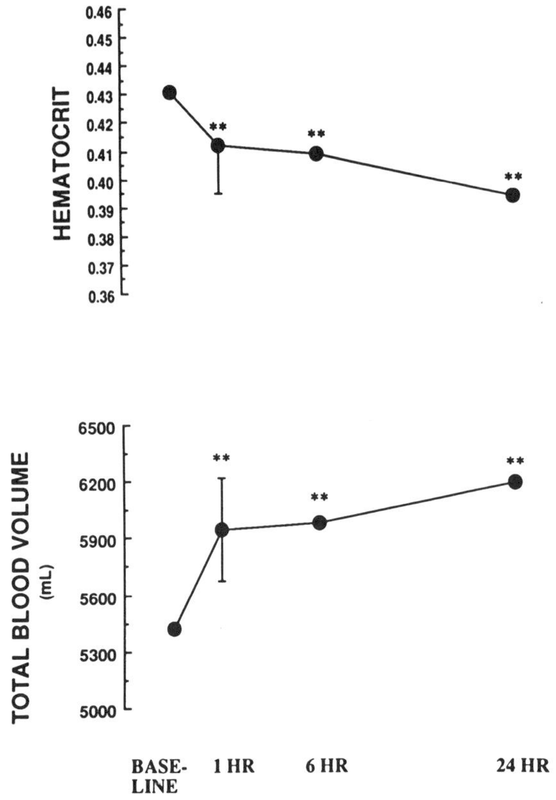

Fig. 3. *Changes in hematocrit and total blood volume during nitroglycerin infusion in 13 patients with chronic congestive heat failure. Hematocrit decreases while total blood volume increases. Body weight and urine output did not vary in these patients so that a capillary fluid shift following a decreased intracapillary pressure must have occurred. **$p < 0.01$ versus control. (Reproduced in part from Dupuis et al. [7], with permission.)*

shown to occur with all vasodilators, also likely plays a role in tolerance [13].

Tolerance as a result of reflex neurohumoral activation is a third potential mechanism. However, studies looking at neurohumoral activation during nitrate therapy have also been conflicting. Some studies found no variation in plasma renin activity, epinephrine, or norepinephrine levels during continuous 24-hour therapy [30,31,40]. Others [13], however, found a significant increase in plasma renin activity after 48 hours of continuous intravenous nitroglycerin infusion with a concomitant increase in body weight, suggesting that neurohumoral activation contributed to tolerance. Unger et al. [38] found no variations in plasma renin activity and norepinephrine levels after 24 hours, but noted a significant elevation in mean plasma epinephrine levels. With intravenous nitroglycerin for 72 hours, Dupuis et al. [6] found transient elevations in plasma renin activity and norepinephrine values 1 hour after the beginning of therapy, a time at which there was maximal hemodynamic effect. The authors failed to find any significant variations in arginine vasopressin, atrial natriuretic peptide, epinephrine, or aldosterone values. Interestingly, neuropeptide-Y values decrease during nitrate therapy [41]. Combining an angiotensin converting enzyme inhibitor with nitrate therapy has been advocated as a means of preventing tolerance by blocking neurohumoral activation and, in the case of captopril, by supplying sulfhydryl groups. However, in two separate studies captopril failed to prevent tolerance in congestive heart failure patients [8,42]. Enalapril was also unsuccessful [38]. Thus the addition of an ACE inhibitor for the sole purpose of tolerance prevention cannot be recommended. While neurohumoral activation may partly contribute to nitrate tolerance in CHF, the above data entail only a slight influence on tolerance.

As for angina patients, there is presently no proven method of preventing tolerance in congestive heart failure with sustained continuous administration. A minimum interval of 8 nitrate-free hours is needed to regain nitrate potency [6] (Figure 4). Providing a nitrate-free interval of at least 8 hours seems to provide maximal benefit during the hours of nitrate utilization [10,13,15].

Exercise Capacity

A major goal of therapy in congestive heart failure is symptomatic improvement. Many trials, some placebo controlled, have demonstrated the beneficial effect of nitrates on exercise performance. In most of these trials, nitrates were administered continuously without regard to a nitrate-free interval. Isosorbide dinitrate in doses of 40 mg four times a day was the regimen utilized most often.

Acutely, nitrates improve submaximal exercise tolerance but not maximal exercise capacity [11,33,34,43]. The inability of heart failure patients to improve maximal exercise capacity despite lower exercise filling pressures is thought to result from peripheral limiting factors not acutely modified by nitrates. With chronic therapy, maximal exercise improves with an increase in maximal exercising time and peak oxygen consumption. After 3 months of chronic isosorbide dinitrate therapy, Leier et al. [34] have elegantly demonstrated a persistent lowering of filling pressures at all levels of exercise. The improvement of maximal exercise performance with chronic therapy is thought to result from a physiologic train-

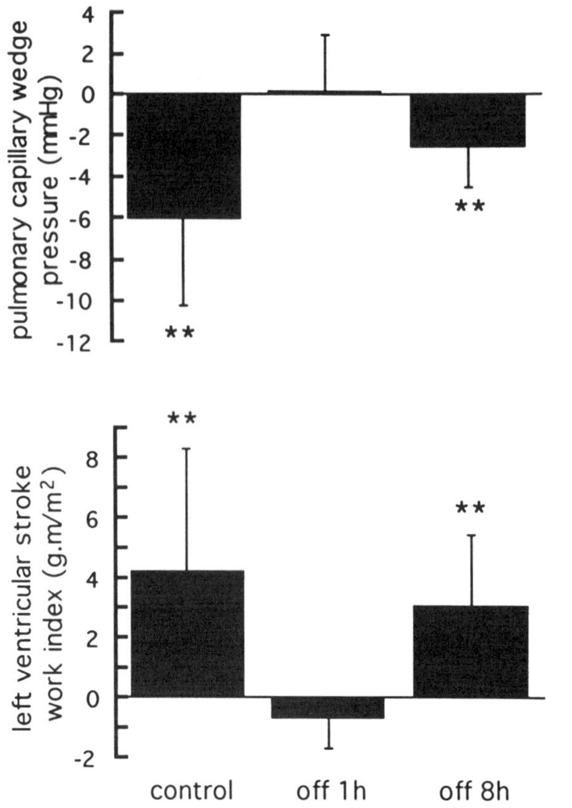

Fig. 4. *Effect of 0.8 mg of sublingual nitroglycerin spray on pulmonary capillary wedge pressure and left ventricular stroke work index. Changes in pressures are shown at baseline and at 1 and 72 hours after withdrawal of a continuous nitroglycerin infusion. There is partial reversal of tolerance after 8 hours of a nitrate-free interval. N = 14. **$p < 0.01$ for a significant change.*

ing effect of nitrates that enables patients to better tolerate longer periods of submaximal exercise in their daily activities. In these trials, nitrates were added to standard heart failure therapy that consisted of digitalis and diuretics. Angiotensin-converting enzyme inhibitors also improve exercise performance [44,45]. Whether nitrate action is additive to that of angiotensin converting enzyme inhibitors remains to be determined.

Mortality

The effects of nitrates as a unique vasodilator therapy on heart failure mortality have never been assessed. Two large randomized trials have, however, evaluated the combination of isosorbide dinitrate and hydralazine (40 mg four times a day and 75 mg four times a day, respectively). In the V-Heft I trial [46], the combination proved superior to placebo or prazosin alone with a 34% 2-year risk reduction in mortality (25.6% mortality for isosorbide dinitrate-hydralazine vs. 34.3% for placebo). The recent V-Heft II trial [47], compared the same combination to angiotensin converting enzyme inhibitor therapy with enalapril (10 mg twice daily). Enalapril slightly but significantly reduced the 2-year mortality when compared with the isosorbide-hydralazine group (18% vs. 25% mortality, respectively, 28% risk reduction). Interestingly, however, left ventricular ejection fraction increased significantly more in the combination therapy group, while peak exercise oxygen consumption improved only in this group.

Clinical Situations

Acute pulmonary edema

Nitrates are extremely effective in rapidly relieving the symptoms of acute pulmonary edema [48]. They acutely reduce all filling pressures, improve cardiac output, reduce mitral regurgitation, and may favor shifts of alveolar fluid to the intravascular space. They should be administered in a manner that will assure the fastest onset of action. When possible, the intravenous route is preferred. We recommend to start the infusion at 0.25 μg/kg/min, with increments of 0.25 μg/kg/min every 5 minutes up to 2 μg/kg/min or until a desired lower blood pressure level has been achieved. Alternately, sublingual nitrates may be administered every 5 minutes, with cumulative doses of up to 2.4 mg being well tolerated.

Refractory heart failure

Patients with severe heart failure refractory to standard medical therapy may benefit from a short course of intravenous nitroglycerin. In 22 patients with class III or IV heart failure [6], there was a sustained beneficial hemodynamic effect during a 72-hour infusion of nitroglycerin in doses of 1.5 μg/kg/min. This beneficial effect seems to be additive to those of dopamine or dobutamine [49–51], so that they may be used concomitantly in patients who need temporary intravenous pharmacologic support.

Chronic stable congestive heart failure

The exact role of nitrates in chronic stable congestive heart failure still needs to be defined. Nitrates can be administered to symptomatic patients with congestive heart failure. Their innocuousness combined with a low cost still make them an agent of choice. Presently, it makes good sense to administer nitrates in a fashion that provides a nitrate-free interval of at least 8 hours so that patients can derive maximal benefits in the hours of activity. Whether intermittent therapy will chronically be more effective than continuous therapy remains to be determined. For patients who do not tolerate angiotensin converting enzyme inhibitors, nitrates can be combined with hydralazine to reduce mortality. Future trials should enable us to define the optimal dosages and posologies for nitrates in congestive heart failure. We also need to delineate the exact role and importance of nitrates in the combined vasodilator therapies of heart failure patients.

Congestive heart failure is associated with vascular endothelial cell dysfunction with reduced endothelium derived relaxing factor production (EDRF) [52] but increased endothelium derived constricting factors such as endothelin [53]. Nitrates are biotransformed in the smooth muscle cell to nitric oxide or a related compound and can thus be considered as an exogenous source of EDRF (alternately, EDRF could be viewed as an endogenous nitrate) whose production is independent of the presence of vascular endothelium. Subsequently, nitrate administration in CHF could be regarded as an attempt to reproduce the normal endothelial cell balance. Although the etiologies, mechanisms, and importance remain to be elucidated [53], the discovery of vascular endothelium dysfunction has brought a new rationale to the use of nitrates in conditions such as congestive heart failure. We now need to rediscover nitrates in that new perspective.

Acknowledgments

The author would like to thank Dr. Jean L. Rouleau for his helpful comments, Dr. Josée Lavoie for her editorial assistance, and Marguerite Cloutier for typing the manuscript.

References

1. Johnson JB, Gross JF, Hale E. Effects of sublingual administration of nitroglycerin on pulmonary artery pressure in patients with failure of the left ventricle. *N Engl J Med* 1957;257:1114.
2. Cohn JN, Franciosa JA. Vasodilator therapy of cardiac failure (first of two parts). *N Engl J Med* 1977;297:27–31.
3. Cohn JN, Franciosa JA. Vasodilator therapy of cardiac failure (second of two parts). *N Engl J Med* 1977;297:254–258.
4. Cohn JN. Nitrates for congestive heart failure. *Am J Cardiol* 1985;56:19A–23A.
5. Packer M. Are nitrates effective in the treatment of chronic heart failure? Antagonist's viewpoint. *Am J Cardiol* 1990; 66:458–461.
6. Dupuis J, Lalonde G, Lebeau R, Bichet D, Rouleau JL. Sustained beneficial effect of a seventy-two hour intravenous infu-

sion of nitroglycerin in patients with severe chronic congestive heart failure. *Am Heart J* 1990;120:625–637.

7. Dupuis J, Lalonde G, Lemieux R, Rouleau JL. Tolerance to intravenous nitroglycerin in patients with congestive heart failure: Role of increased intravascular volume, neurohumoral activation and lack of prevention with n-acetylcysteine. *J Am Coll Cardiol* 1990;16:923–931.
8. Dupuis J, Lalonde G, Bichet D, Rouleau JL. Captopril does not prevent nitroglycerin tolerance in heart failure. *Can J Cardiol* 1990;6:281–286.
9. Leier CV, Bambach D, Thompson MJ, Cattaneo SM, Goldberg RJ, Unverferth DV. Central and regional hemodynamic effects of intravenous isosorbide dinitrate, nitroglycerin and nitroprusside in patients with congestive heart failure. *Am J Cardiol* 1981;48:1115–1123.
10. Sharpe N, Coxon R, Webster M, Luke R. Hemodynamic effects of intermittent transdermal nitroglycerin in chronic congestive heart failure. *Am J Cardiol* 1987;59:895–899.
11. Franciosa JA, Nordstrom LA, Cohn JN. Nitrate therapy for congestive heart failure. *JAMA* 1978;240:443–446.
12. Franciosa JA, Blank RC, Cohn JN. Nitrate effects on cardiac output and left ventricular outflow resistance in chronic congestive heart failure. *Am J Med* 1978;64:207–213.
13. Packer M, Lee WH, Kessler PD, Gottlieb SS, Medina N, Yushak M. Prevention and reversal of nitrate tolerance in patients with congestive heart failure. *N Engl J Med* 1987;317:799–804.
14. Armstrong PW, Armstrong JA, Marks GS. Pharmacokinetic-hemodynamic studies of nitroglycerin ointment in congestive heart failure. *Am J Cardiol* 1980;46:670–676.
15. Elkayam U, Roth A, Mehra A, et al. Randomized study to evaluate the relation between oral isosorbide dinitrate dosing interval and the development of early tolerance to its effect on left ventricular filling pressure in patients with chronic heart failure. *Circulation* 1991;84:2040–2048.
16. Williams DO, Bommer WJ, Miller RR, Amsterdam EA, Mason DT. Hemodynamic assessment of oral peripheral vasodilator therapy in chronic congestive heart failure: Prolonged effectiveness of isosorbide dinitrate. *Am J Cardiol* 1977;39:84–90.
17. Bayley S, Valentine H, Bennett ED. The haemodynamic responses to incremental doses of intravenous nitroglycerin in left ventricular failure. *Intensive Care Med* 1984;10:139–145.
18. Jordan RA, Seth L, Henry A, Wilen MM, Franciosa JA. Dose requirements and hemodynamic effects of transdermal nitroglycerin compared with placebo in patients with congestive heart failure. *Circulation* 1985;71:980–986.
19. Packer M, Medina N, Yushak M, Lee WH. Hemodynamic factors limiting the response to transdermal nitroglycerin in severe chronic congestive heart failure. *Am J Cardiol* 1986;57:260–267.
20. Rajfer SI, Demma FJ, Goldberg LI. Sustained beneficial hemodynamic responses to large doses of transdermal nitroglycerin in congestive heart failure and comparison with intravenous nitroglycerin. *Am J Cardiol* 1984;54:120–125.
21. Packer M. Mechanisms of nitrate action in patients with severe left ventricular failure: Conceptual problems with the theory of venosequestration. *Am Heart J* 1985;110:259–264.
22. Keren G, Bier A, Strom JA, Laniado S, Sonnenblick EH, Lejemtel TH. Dynamic of mitral regurgitation during nitroglycerin therapy: A Doppler echocardiographic study. *Am Heart J* 1986;112:517–525.
23. Keren G, Katz S, Strom J, Sonnenblick E, LeJemtel TH. Dynamic mitral regurgitation. An important determinant of the

hemodynamic response to load alterations and inotropic therapy in severe heart failure. *Circulation* 1989;80:306–313.
24. Cohn JN. Role of nitrates in congestive heart failure. *Am J Cardiol* 1987;60:39H–43H.
25. Schneider AJ, Teule GJ, Groeneveld AB, Nauta J, Luth WJ. The immediate effect of nitroglycerin on total body volume distribution in patients with congestive heart failure: A non-invasive study. *Eur Heart J* 1987;8:1119–1125.
26. Smith ER, Smiseth OA, Kingma I, Manyari D, Belenkie I, Tyberg J. Mechanism of action of nitrates. Role of changes in venous capacitance and in the left ventricular diastolic pressure-volume relation. *Am J Med* 1984;76:14–21.
27. Winbury MM. Redistribution of left ventricular blood flow produced by nitroglycerin. *Circ Res* 1971;28:140–147.
28. Gorman MW, Sparks HV. Nitroglycerin causes vasodilation within ischemic myocardium. *Cardiovasc Res* 1980;14:515.
29. Greenberg H, Dwyer EM, Jameson AG, et al. Effects of nitroglycerin on the major determinants of myocardial oxygen consumption. An angiographic and hemodynamic assessment. *Am J Cardiol* 1975;36:426–432.
30. Elkayam U, Kulick D, McIntosh N, Roth A, Hsueh W, Rahimtoola SH. Incidence of early tolerance to hemodynamic effects of continuous infusion of nitroglycerin in patients with coronary artery disease and heart failure. *Circulation* 1987;76:577–584.
31. Olivari MI, Carlyle PF, Levine TB, Cohn JN. Hemodynamic and hormonal response to transdermal nitroglycerin in normal subjects and in patients with congestive heart failure. *J Am Coll Cardiol* 1983;2:872–878.
32. Sharpe N, Coxon R. Nitroglycerin in a transdermal therapeutic system in chronic heart failure. *J Cardiovasc Pharmacol* 1984;6:76–82.
33. Franciosa JA, Cohn JN. Effect of isosorbide dinitrate on response to submaximal and maximal exercise in patients with congestive heart failure. *Am J Cardiol* 1979;43:1009–1014.
34. Leier CV, Huss P, Magouin RD, Unverferth DV. Improved exercise capacity and differing arterial and venous tolerance during chronic isosorbide dinitrate therapy for congestive heart failure. *Circulation* 1983;67:817–822.
35. Franciosa JA, Cohn JN. Sustained hemodynamic effects without tolerance during long term isosorbide dinitrate treatment of chronic left ventricular failure. *Am J Cardiol* 1980;45:648–654.
36. Jordan RA, Seth L, Casebolt P, Hayes MJ, Wilen MM, Franciosa J. Rapidly developing tolerance to transdermal nitroglycerin in congestive heart failure. *Ann Intern Med* 1986;104:295–298.
37. Münzel T, Holtz J, Mülsch A, Stewart DJ, Bassenge E. Nitrate tolerance in epicardial arteries O_2 in the venous system is not reversed by N-acetylcysteine in vivo, but tolerance-independent interactions exist. *Circulation* 1989;79:188–197.
38. Unger P, Leone A, Staroukine M, Degre S, Berkenboom G. Hemodynamic response to molsidomine in patients with ischemic cardiomyopathy tolerant to isosorbide dinitrate. *J Cardiovasc Pharmacol* 1991;18:888–894.
39. Parker JD, Farrel B, Parker AC, Cahanim MA, Parker JO. Neurohumoral responses and the hemodynamic adaptation to nitrates (abstr). *Circulation* 1990;82(Suppl III):III200.
40. Elkayam U, Roth A, Henriquez B, Weber L, Tonnemacher D, Rahimtoola SH. Hemodynamic and hormonal effects of high-dose transdermal nitroglycerin in patients with chronic congestive heart failure. *Am J Cardiol* 1985;56:555–559.
41. Derchi G, Dupuis J, de Champlain J, Rouleau JL. Paradoxical decrease in circulating neuropeptide-Y levels during mild

orthostatic stress in normal and patients with congestive heart failure. *Eur Heart J* 1992;14:34–39.

42. Dakak N, Makhoul N, Fugelman MY, et al. Failure of captopril to prevent nitrate tolerance in congestive heart failure secondary to coronary artery disease. *Am J Cardiol* 1990;66:608–613.
43. Wilson JR, Ferraro N. Effect of isosorbide dinitrate on submaximal exercise capacity of patients with chronic left ventricular failure. *Chest* 1982;82:701–704.
44. Creager MA, Massie BM, Faxon DP, et al. Acute and long term effect of enalapril on the cardiovascular response to exercise in patients with congestive heart failure. *J Am Coll Cardiol* 1985;6:163–173.
45. Packer M, Medina N, Yushak M. Comparative hemodynamic and clinical effect of long-term treatment with prazosin and captopril for severe chronic congestive heart failure secondary to coronary artery disease or idiopathic dilated cardiomyopathy. *Am J Cardiol* 1986;57:1323–1327.
46. Cohn JN, Archibald DG, Ziesche S, et al. Effect of vasodilator therapy on mortality of chronic congestive heart failure. *N Engl J Med* 1986;314:1547–1552.
47. Cohn JN, Johnson G, Ziesche S, et al. A comparison of enalapril with hydralazine-isosorbide dinitrate in the treatment of chronic congestive heart failure. *N Engl J Med* 1991;325:303–310.
48. Bussmann WD, Schupp D. Effect of sublingual nitroglycerin in emergency treatment of severe pulmonary edema. *Am J Cardiol* 1978;41:931–936.
49. Loeb H, Ostrenga JP, Gaul W, et al. Beneficial effects of dopamine combined with intravenous nitroglycerin on hemodynamics in patients with severe left ventricular failure. *Circulation* 1983;68:813–820.
50. Awan NA, Evenson NK, Needham KE, Beattie JM, Mason DT. Effect of combined nitroglycerin and dobutamine infusion in left ventricular dysfunction. *Am Heart J* 1983;106:35–40.
51. Gagnon RM, Fortin L, Boucher R, et al. Combined hemodynamic effects of dobutamine and IV nitroglycerin in congestive heart failure. *Chest* 1980;78:694–698.
52. Kubo HS, Rector TS, Bank AJ, Williams RE, Heifetz SM. Endothelium-dependent vasodilation is attenuated in patients with heart failure. *Circulation* 1991;84:1589–1596.
53. Stewart DJ, Cernacek P, Costello KB, Rouleau JL. Elevated endothelin-1 in heart failure and loss of normal response to postural change. *Circulation* 1992;85:510–517.
54. Treasure CB, Alexander RW. The dysfunctional endothelium in heart failure. *J Am Coll Cardiol* 1993;22(Suppl A): 129A–134A.

The Organic Nitrates: Practical Problems In Their Use

Jonathan Abrams

Summary. **Organic nitrate adverse effects directly relate to their hemodynamic actions: vasodilatation and reflex sympathetic discharge. Thus, hypotension may induce dizziness or presyncope. Cerebral vasodilatation produces headaches. Nitrate tolerance is probably multifactorial, and results in attenuation or loss of nitrate action. Tolerance is well documented in angina pectoris and congestive heart failure. The only proven strategy to prevent tolerance is to employ a nitrate-free dosing interval, and avoid sustained nitrate delivery.**

Key Words. **isosorbide dinitrate, nitroglycerin, tolerance, resistance, nitrate-free interval**

Nitroglycerin (NTG) and the long acting nitrates have been used in clinical practice for over 100 years. Their track record attests to the safety and ease with which these drugs can be administered. The side effect profile of the nitrates is limited and well recognized, consisting mainly of headache and occasional hypotensive symptoms. Reports of other problems appear in the literature, but are uncommon and usually inconsequential. A substantial number of individuals will not take NTG or long acting nitrates because of nitrate-induced headaches; data are not available to confirm the magnitude of this problem. Drug interactions are rare, although hypotension may be exacerbated when nitrates are administered concomitantly with other potent vasodilators or in the setting of volume depletion.

The major problem with long acting nitrate therapy is nitrate tolerance. Attenuation of effects or tolerance to nitrates means that there are diminishing pharmacologic and hemodynamic responses during continued administration of the drug in the absence of alterations in bioavailability or decreased plasma or tissue concentration. Extensive research continues today in efforts to understand and ultimately overcome this vexing problem. Finally, nitrate resistance, a little known problem, is another phenomenon that may be important in patients with severe congestive heart failure receiving nitrates for vasodilator therapy. Resistance has not been shown to be a serious problem in the treatment of angina pectoris, although there are virtually no available data.

This chapter will discuss the major practical problems in nitrate therapy, emphasizing nitrate tolerance and resistance.

Acute Side Effects

Most of the adverse reactions due to nitrates appear immediately and are directly related to the vasodilating properties of these drugs. *Headache*, or a sensation of throbbing or pounding of the head and neck, is common, and most individuals who use sublingual nitroglycerin (NTG) have experienced symptoms related to this phenomenon. It is presumed that the headache is caused by an increase in intracranial pressure due to intracerebral arterial vasodilatation. However, there are little or no data concerning headaches or the responsible vascular bed. The headache can range from excruciatingly severe, accompanied by systemic symptoms of nausea and vomiting, to a mild discomfort or head throbbing. Little is known about the time course of headache development and disappearance; it appears that the duration of headaches parallels the known vasodilatory actions and pharmacokinetic activity of the administered nitrate. Thus, for sublingual or oral spray NTG, the headache will last no more than 15 to 30 minutes. However, individuals who use NTG patches may experience continuing symptoms ranging from very mild to intolerance. Headaches usually diminish over time with continuing nitrate administration, and in many individuals, ultimately disappear completely. Thus, physicians should counsel patients regarding this problem, explaining that it is best to try to get through the first few days or weeks of long acting nitrate therapy. The use of acetaminophen or aspirin appears to help, especially when administered 30 minutes prior to scheduled doses, and may be useful in tiding patients through the early period when headaches are the most intense. However, it is also apparent from clinical practice that a substantial number of patients, perhaps 15 to 30%, will not be able to tolerate nitrates given acutely or during long acting therapy, and therefore, other medication must be used. One recent study suggested that ice supplied to the neck may be useful in diminishing the headaches [1].

Hypotension, manifest by dizziness, lightheadedness, or occasionally overt syncope, is a well recognized adverse response to nitrate therapy, and is caused by the arterial and venous vasodilating actions of these drugs. Nitrates are more potent in the upright position, which results in a greater fall in systolic arterial pressure, as well as enhanced venodilatation; cardiac chamber size is smaller and preload is further reduced in the standing or sitting position. Thus, individuals who are standing are more likely to experience hypotensive symptoms. Furthermore, we have shown that alcohol acts additively to nitrate hypotension (and reflex tachycardia), and therefore, individuals who take sublingual or spray NTG should be warned of potential side effects [2].

Some patients are exquisitely sensitive to NTG and may experience presyncope or even faint with the first dose. Therefore, physicians should always *initiate therapy with low doses*, building up to an effective dose, whether it be with short or long-acting nitrates. I usually suggest that patients take their first "test" dose at home when they are angina-free, so as to become familiar with the effects of NTG. Tolerance to the immediate hypotensive effects of nitrates occurs with continued nitrate administration, although this has not been well studied. In the presence of hypovolemia (excessive diuresis, dehydration), low baseline blood pressure, right ventricular infarction, or concurrent administration of other vasoactive drugs, nitrates may be particularly problematic by producing a symptomatic fall in arterial pressure. Patients should be advised to sit or lay down immediately if lightheadedness occurs following sublingual NTG or long acting nitrate administration. Elevation of the legs, accompanied by assumption of the supine position, is usually all that is necessary to rapidly reverse symptomatic hypotension. Occasionally, volume administration is useful in hospitalized patients.

Patients with obstruction of the left ventricular outflow are typically preload dependent. NTG can precipitate hypotension in both valvular aortic stenosis and hypertrophic cardiomyopathy, and in the latter case, can also potentiate the degree of obstruction by lowering afterload.

Palpitations and *tachycardia* commonly follow nitrate administration, particularly sublingual or aerosol spray NTG. This is due to reflex sympathetic discharge; heart rate may accelerate remarkably, especially in patients who have a pronounced sympathetic discharge. Cases of nitrate-induced *bradyarrhythmias* including heart block have been described [3,4], usually associated with hypotension and marked vagotonia. Thus, patients who experience lightheadedness, presyncope, or overt syncope following nitrates, may suffer from bradycardia as well as marked peripheral vasodilatation to a degree out of proportion to the hemodynamic effects of the drug itself.

Miscellaneous Problems With Nitrates

In the early nitrate literature, it was recommended that these drugs not be used in patients with glaucoma; however, in chronic open angle glaucoma, the most common form, has been shown to be improved by nitrate administration [5]; effects in narrow angle glaucoma, have not been studied. Nitrates should not be administered to patients with increased intracranial pressure [6]. High dose, prolonged intravenous nitroglycerin administration can produce methemoglobinemia [7], particularly in subjects with glucose-3-phosphodiesterase (G3PD) deficiency, but this is a rare event. Recent data suggests that nitrates lessen the effect of tissue plasminogen (tPA) activator [8]. The clinical significance of these observations remain to be explored. Nitrate-induced arterial desaturation has been documented in chronic lung disease patients, presumably due to increased pulmonary A-V shunting. This is more of a theoretical than practical concern, however.

Drug interactions

Few specific drug interactions have been identified for the nitrates. As previously suggested, concomitant nitrate administration with calcium channel blockers, angiotensin converting enzyme (ACE) inhibitors, morphine sulfate or diuretics, may excessively lower arterial pressure. The combination of a dihydropyridine calcium antagonist and a nitrate is generally to be avoided because of the hypotensive potential of two potent arterial vasodilating drugs. Beta blocker administration does not alter nitrate pharmacokinetics [9]. Concomitant administration of intravenous NTG and heparin has been associated with nitrate-induced shortening of the activated partial thromboplastin time (PTT); this interaction, while of potentially great importance in the intensive care unit, remains controversial [10,11]. Physicians should use caution when starting or stopping intravenous NTG in patients on an intravenous heparin drip.

Hepatic and renal disease

Limited data suggest that there is no alteration in nitrate pharmacokinetics in the presence of significant renal or hepatic disease [9]. Nevertheless, because of the extensive hepatic metabolism of certain orally administered nitrates, patients with cirrhosis and impaired liver function should be given oral ISDN or NTG with caution.

Nitroglycerin patch problems

Unusual reactions to transdermal NTG patches have been reported. It is not uncommon to have minor skin irritation, manifest by erythema, and occasionally more pronounced irritative phenomena such as blistering, usually due to the adhesive. When placed on the chest, the patches have been rarely associated with electrical surface burns during defibrillation or elective cardioversion.

Intravenous NTG problems

NTG is readily adsorbed onto conventional polyvinylchloride (PVC) tubing, and the delivered concentration may be less than that desired. It has been suggested that specialized infusion sets be utilized in the patients receiving intravenous NTG. This is probably not necessary nor worth the expense. Another potential problem relates to the alcohol content in commercially prepared intravenous NTG solutions; patients who receive high dose infusion for protracted time periods could become intoxicated, particularly if they are taking disulfiram-like drugs (e.g. Antabuse, Flagyl, certain third generation cephalosporins). Some NTG preparations have a high potassium content and can cause ventricular fibrillation when administered intracoronary.

Nitrate Tolerance

Beginning around 1980 an increasing number of high quality clinical investigations into nitrate tolerance were published, convincing the medical world that this was indeed a valid issue [12–17]. Prior to that time, most clinicians did not believe that tolerance was a real problem in nitrate therapy. In fact, it remains difficult for a physician to identify a patient in whom nitrate tolerance is clearly present. The discrepancy between the clinical experience with nitrate therapy, relating to clinical stability of patients taking long acting nitrates and their continued therapeutic responsiveness to sublingual NTG, and the research data confirming that attenuation to nitrate action readily appears, is not well explained. Nevertheless, the work of Parker, Thadani, Rudolph, and scores of others have firmly placed this issue on the agenda of clinicians and investigators who wish to administer nitrates.

What has emerged over the past decade or more began as disparate pieces of data that now form a clear story. While it remains unclear as to the actual mechanism(s) of nitrate tolerance, it is well recognized how tolerance can be induced and/or prevented. *The only practical and effective way to avoid or diminish this problem is to use a nitrate dosing strategy that maximizes vascular responsiveness and minimizes attenuation of vasodilation.* Quite simply, the greater the exposure of veins and arteries to nitrates, the more readily tolerance is induced, and the more severe the degree of tolerance. Conversely, less nitrate exposure is less likely to result in attenuation of nitrate effects. The goal is to administer enough nitrate to produce the desired clinical benefits while avoiding the induction of the tolerant state.

Appropriate dosing regimens can provide continuing nitrate efficacy for days, weeks, or months. The obvious downside, however, to such nitrate regimens, is the inability to provide around the clock nitrate action because of the nitrate-free or low period that must be included in the dosing recommendation. Thus, in severely symptomatic patients with angina or heart failure, additional pharmacologic agents are usually needed to assure therapeutic effects during the nitrate free period.

Consequences of Nitrate Tolerance

The reduction of nitrate vasodilating effects lead to a predictable loss of therapeutic efficacy, depending on the clinical condition for which the nitrates are given. Thus, in angina pectoris, nitrate tolerance is associated with diminished anti-ischemic efficacy, manifested by objective evidence of a shorter time to ST segment depression, as well as to anginal discomfort. Exercise time to the onset of chest pain or total exercise duration typically returns to baseline or placebo values or is diminished from the maximal nitrate effect in the presence of nitrate tolerance [12–13,18–20]. For instance, it has been consistently shown that isosorbide dinitrate (ISDN), when given four times a day or every six hours, reduces the total duration of nitrate action from 4 to 6 hours to 2 to 3 hours, and may also decrease the peak effect of nitrates [12,19,20]. Other studies have shown a complete loss of antianginal effect when groups of patients receiving protolerant nitrate regimens are studied in placebo controlled trials [18,21–22]. On the other hand, there are limited data that document a loss of subjective or clinical responses to long acting nitrate therapy. Thus, patients who have been documented to have tolerance by exercise testing do not typically report an increase in angina attack rates or sublingual nitroglycerin consumption. This discrepancy is unexplained.

In congestive heart failure, nitrate tolerance has been shown to appear rapidly, often within 12 to 24 hours after initial administration, and is associated with attenuation of nitrate effects on pulmonary capillary wedge pressure, pulmonary artery, and mean arterial pressure [14–17,23–25]; concomitantly, stroke volume responses are blunted when compared to acute nitrate administration [24–25].

It is important to emphasize that tolerance studies typically provide group data, that is, mean responses of the population under study; relatively few investigations also report individual patient responses. Some investigators have found that tolerance does not appear in every patient, and that in fact, some individuals will maintain complete responsiveness to nitrate administration although the entire patient cohort may demonstrate tolerance [24]. *Thus, it is important to recognize that tolerance is not necessarily an all-or-none phenomenon, nor does it occur in every individual.* Furthermore, there is evidence that tolerance may occur more readily in the venous bed than in the arterial circulation [26]; it is possible that different vascular tissues respond differently to nitrates, perhaps relating to differing vascular smooth muscle uptake of organic nitrates [27] and the ease or resistance to induction of tolerance. In addition, it is likely that tolerance mechanisms, time course, and severity are different in angina from congestive heart failure.

There is no reason to believe that tolerance is less or more of a problem in angina, heart failure, or even in acute myocardial infarction when continuous or frequent nitrate administration is employed. Furthermore, there is no clear evidence that one nitrate molecule is more tolerance promoting than another. Tolerance is directly related to high or sustained plasma concentrations, the presence of absence of an "adequate" nitrate free interval, and possibly the pharmacokinetic profile of the administered nitrate. In anginal patients, efforts to overcome tolerance with enormous doses of nitrates have been unsuccessful, in spite of some preliminary studies to the contrary [28].

Strategies to Avoid Nitrate Tolerance

The approaches used by investigators in efforts to overcome or prevent the development of tolerance are related to the various hypotheses of how and why nitrate tolerance occurs. Efforts to provide thiol groups to replenish or replete sulfhydryl group deficiency have been given a great deal of attention [29]. Similarly, attempts to blunt or interfere with the neurohumoral activation that occurs with nitrate administration and may be related in part to attenuation of nitrate effects, has been given considerable attention. In this regard, angiotensin converting enzyme inhibitors and diuretics have been administered in efforts to counteract nitrate-induced increases in renin-angiotensin and sympathetic nervous system activity. At present, none of those strategies can be recommended.

The only practical way to prevent or decrease the likelihood of tolerance is to employ dosing strategies that empirically have been proven to maintain nitrate responsiveness [30–35]. In general, these include less frequent doses, lower plasma nitrate concentrations, and provision of a prolonged nitrate free interval on a daily basis (Table 1). For whatever reasons, the cessation of nitrate exposure to vascular smooth muscle allows for restoration of full vasodilatory capacity within several hours. The required nitrate free period probably varies among vascular tissues and various nitrate compounds, and is clearly dose and exposure time related. It has been suggested that, ideally, plasma nitrate concentrations should rise and fall throughout the day [35], as opposed to maintenance of a steady state nitrate input and sustained plasma concentrations. Experience with NTG patches and intravenous NTG indicate that it is easier to induce tolerance with these formulations than with intermittent or pulsed therapy with oral nitrates. In any case, the concept that nitrate plasma and tissue concentrations must fall to a low level each day for a prolonged period of time has been validated by numerous clini-

Table 1. *Pharmacologic guidelines for avoiding nitrate tolerance*

- Minimize number of doses per day
- Provide a nitrate-free or low interval each day. This interval varies with the preparations used (31b & 31c)

Congestive heart failure

- Larger than usual doses are usually required in congestive heart failure

cal trials, whether intermittent or continued nitrate dosing formulations are utilized.

Problems With the Nitrate-Free Interval

It has been argued that the necessity for a nitrate-free interval and the necessary inability to provide active nitrate exposure for a portion of the day puts patients at risk. Thus, in individuals with ischemic heart disease and angina pectoris, it is possible that there will be increased potential for myocardial ischemic events during the nitrate-free period. In congestive heart failure, there will be a return toward baseline levels of hemodynamic parameters as nitrate levels decline. There is no way to avoid these problems. For patients with significant ischemia or frequent angina, additional antianginal agents are required. In heart failure, diuretics and ACE inhibitor therapy in conjunction with nitrates appear to provide sufficient protection to the patient. Rebound vasoconstrictor forces may also play a role during the nitrate-free period, but this remains controversial (see below).

Safety of the Nitrate-free Interval

Several reports have suggested that there may be an increase in ischemic events during the nitrate-free interval during intermittent NTG patch administration [36–37]. However, several other reports have been unable to confirm this problem [38–39]; at the present time, one should maintain a cautionary position with respect to the possibility of patch-off angina. Physicians should alert their patients to report any new nocturnal symptoms during the patch-off period. Similar problems have been suggested to occur during oral nitrate therapy [40].

The problem of *"patch-off" angina* may relate to rebound coronary and systemic vasoconstriction following patch removal. This may be comparable to nitrate dependence that has been well documented among nitroglycerin manufacturing workers [41]. Coronary vasoconstriction and overt coronary artery spasm have been suggested as potentially hazardous sequelae of withdrawal from the NTG laden work environment; such phenomena might also occur when a patch is removed after prolonged application.

Nitrate Resistance

A little known problem of nitrate therapy that has been documented on occasion over the past decade describes the inability of veins and arteries to respond (e.g., dilate) when nitrates are given to patients with congestive heart failure [41–45]. When one carefully evaluates the literature, it is clear that patients who are likely to be resistant to administration of nitrates are those with relatively severe heart failure, often with clinical evidence of biventricular failure. While this is not always the case, several studies have indicated that a relatively high right atrial pressure and/or peripheral edema are markers for resistance, and that diuretic therapy may restore vascular responsiveness [42]. It is unclear if nitrate resistance is an accelerated or exaggerated variant of nitrate tolerance, nor are the mechanisms fully understood. In general, patients with heart failure require larger nitrate doses than conventionally used in angina pectoris. To overcome nitrate resistance, it has been documented that very large doses may be necessary in some individuals [45].

Elkayam and colleagues have done the most careful studies of this problem, and have been able to overcome resistance in a substantial number of patients who initially did not respond to 40 mg. of isosorbide dinitrate; some of these individuals ultimately required 80 or 120 mg. of ISDN per dose, and even at these levels occasional patients remain resistant [45].

Conclusion

NTG and the other nitrates have a relatively small number of adverse effects, of which headache is the most common. Drug interactions are few. Except for the potential of inducing hypotension, nitrates are quite safe and can be used safely in patients with impaired left ventricular function. Tolerance is the major limiting factor in nitrate therapy. Carefully designed dosing regimens can prevent most clinically important nitrate attenuation.

Acknowledgements

I would like to thank Michael Reed, Pharm. D., for reviewing the text.

References

1. Broberg DC, Belknap D, Davidson L, Hamilton S. Icepacks for nitroglycerin headaches. *Circulation* 1993;86(Suppl I):1–105.
2. Abrams J, Schroeder K, Raizada V, Gibbs D. Potentially adverse effects of sublingual nitroglycerin dosing consumption of alcohol. *J Am Coll Cardiol* 1990;15:226A.
3. Come PC, Pitt B. Nitroglycerin-induced severe hypotension and bradycardia in patients with acute myocardial infarction. *Circulation* 1976;54:624–628.

4. Khan AH, Carleton RA. Nitroglycerin-induced hypotension and bradycardia. *Arch Intern Med* 1981;141:984–986.
5. Wizemann A, Wizemann V, Krey H. Organic nitrates—a new principle in glaucoma therapy? In Lichtlen PR, Engel HL, Schrey A, and Swan HJC, (eds) Nitrates III. *Cardiovascular Effects*. Berlin: Springer-Verlag, 1981, pp 582–586.
6. Ohar JM, Fowler AA, Selhorst JB, Glaser FL. Intravenous nitroglycerin-induced intracranial hypertension. *Crit Care Med* 1985;13:867–868.
7. Arsura E, Lichstein E, Guadagnino V, et al. Methemoglobin levels produced by organic nitrates in patients with coronary artery disease. *J Clin Pharmacol* 1984;24:160–167.
8. Mehta JL, Nicolini F, Nichols WW, Saldeen TGP, Wilson AL, Thompson LV. Concurrent nitroglycerin administration decreases thrombolytic potential of tissue-type plasminogen activator. *J Am Coll Cardiol* 1991;17:805–811.
9. Bogaert MG, Rosseel MT. Fate of orally given isosorbide dinitrate in man: Factors of variability. *Z Kardiol* 1983;72(Suppl 3):11–13.
10. Habbab M, Haft J. Heparin resistance induced by intravenous nitroglycerin: A word of caution when both drugs are used concomitantly. *Arch Intern Med* 1987;147:856–862.
11. Bode V, Weizel D, Franz G, et al. Absence of drug interaction between heparin and nitroglycerin. Randomized placebo-controlled crossover study. *Arch Intern Med* 1990;150:2117–2119.
12. Thadani U, Fung HD, Darke AC, Parker JO. Oral isosorbide dinitrate in angina pectoris. Comparison of duration of action and dose response relationship during acute and sustained therapy. *Am J Cardiol* 1982;49:411–417.
13. Tauchert M, Jansen W, Osterpey A, Fuchs M, Homback V, Hilger HH. Dose dependence of tolerance during treatment with mononitrates. *Z Kardiol* 1983;72(Suppl 3):218–228.
14. Rudolph W, Blasini R, Froer KL. Blume R, Rudolph W. Wirkungverlust von isosorbiddnitrat bei langzeitbehandlung der chronischen herzinsuffizienz. *Herz* 1982;7:250–258.
15. Blasini R, Froer KL. Blume R, Rudolph W. Wirkungverlust von isosorbiddnitrat bei langzeitbehandlung der chronischen herzinsuffizienz. *Herz* 1982;7:250–258.
16. Leier CV, Huss P, Magorien RD, Unverferth DV. Improved exercise capacity and differing arterial and venous tolerance during chronic isosorbide dinitrate therapy for congestive heart failure. *Circulation* 1983;67:817–822.
17. Olivari MT, Carlyle PF, Levine TB, Cohn JN. Hemodynamic and hormonal response to transdermal nitroglycerin in normal subjects and in patients with congestive heart failure. *J Am Coll Cardiol* 1983;2:872–878.
18. Reichek N, Priest C, Zimrin D, et al. Limited antianginal effects of nitroglycerin patches. *Am J Cardiol* 1984;54:1–6.
19. Parker JO, Vankoughnett KA, Farrell B. Comparison of buccal nitroglycerin and oral isosorbide dinitrate for nitrate tolerance in stable angina pectoris. *Am J Cardiol* 1985;56:724–728.
20. Parker JO, Farrell B, Lahey KA, et al. Effect of intervals between doses on the development of tolerance to isosorbide dinitrate. *N Engl J Med* 1987;316:1440–1444.
21. Parker JO, Fung HL. Transdermal nitroglycerin in angina pectoris. *Am J Cardiol* 1984;54:471–476.
22. Steering Commitee, Transdermal Nitroglycerin Cooperative Study. Acute and chronic antianginal efficacy of continuous

twenty-four-hour application of transdermal nitroglycerin. *Am J Cardiol* 1991;68:1263–1273.

23. Packer M, Lee WH, Kessler PD, et al. Prevention and reversal of nitrate tolerance in patients with congestive heart failure. *N Engl J Med* 1987;317:799–804.
24. Elkayam U, Kulick D, McIntosh N, Roth A, Hsueh W, Rahimtoola SH. Incidence of early tolerance to hemodynamic effects of continuous infusion of nitroglycerin in patients with coronary artery disease and heart failure. *Circulation* 1987;76:577–584.
25. Dupuis J, Lalonde G, Lemieux R, Rouleau JL. Tolerance to intravenous nitroglycerin in patients with congestive heart failure: Role of increased intravascular volume, neurohormonal activation and lack of prevention with N-acetylcysteine. *J Am Coll Cardiol* 1990;16:923–931.
26. Stewart DJ, Elsner D, Sommer O, et al. Altered spectrum of nitroglycerin action in long-term treatment: Nitroglycerin-specific venous tolerance with maintenance of arterial vasodepressor potency. *Circulation* 1986;75:573–579.
27. Fung HL, Sutton SC, Kamiya A. Blood vessel uptake and metabolism of organic nitrates in the rat. *Pharmacol Exp Ther* 1984;228:334–341.
28. Schneider WU, Bussman WD, Stahl B, Kaltenbach M. Dose response relation of antianginal activity of isosorbide dinitrate. *Am J Cardiol* 1984;53:700–705.
29. Abrams J. Interactions between organic nitrates and thiol groups. *Am J Med* 1991;91(Suppl C):106–112.
30. Abrams J. Tolerance to organic nitrates. *Circulation* 1986;74:1181–1185.

31a. Thadani U. Current status of nitrates in angina pectoris. *Mod Con Cardiovasc Dis* 1987;56:49–54.

31b. Thadani U, Lipicky RJ. Short and long acting oral nitrates for stable angina pectoris. *Cardiovasc Druas Ther* 1994;8:611–623.

31c. Thadani U, Lipicky RJ. Ointments and transdermal nitroglycerin patches for stable angina pectoris. *Cardiovasc Drugs Ther* 1994;8:625–633.

32. Elkayam U. Tolerance to organic nitrates: Evidence, mechanisms, clinical relevance, and strategies for prevention. *Ann Intern Med* 1991;114:667–677.
33. Parker JO. Nitrate tolerance. *Am J Cardiol* 1987;60:44H–48H.
34. Abrams J. Clinical aspects of nitrate tolerance. *Eur Heart J* 1991;12(Suppl):42–52.
35. Silber S, Vogler AC, Krause KH, et al. Induction and circumvention of nitrate tolerance applying different dosage intervals. *Am J Med* 1987;83:860–870.
36. DeMots H, Glasser SP. Intermittent transdermal nitroglycerin theapy in the treatment of chronic stable angina. *J Am Coll Cardiol* 1989;13:786–788.
37. Ferrantini M, Pirelli S, Merlini P, Silva T, Pollavini G. Intermittent transdermal nitroglycerin monotherapy in stable exercise induced angina: A comparison with a continuous schedule. *Eur Heart J* 1989;10:998–1002.
38. Fox KM, Dargie H, Dearfield J, Maseri A. The avoidance of tolerance and lack of rebound using intermittent dose titrated transdermal glyceryl trinitrate. *Eur Heart J.*
39. Gumbrielle T, Freedman SB, Fogarty L, Ogasawara S, Sobb P, Kelly DT. Efficacy, safety and duration of nitrate-free interval to prevent tolerance to transdermal nitroglycerin in effort angina. *Eur Heart J* 1992;13:671–678.
40. Rehnquist N, Olsson G, Rosenquist U, et al. Abrupt withdrawal of isosorbide-5-mononitrte in Durules (Imdur) after long term

treatment in patients with stable angina pectoris. *Eur Heart J* 1988;9:1335–1347.
41. Abrams J. Nitroglycerin tolerance and dependence. *Am Heart J* 1980;99:113–123.
42. Magrini F, Niardros AP. Ineffectiveness of sublingual nitroglycerin in acute left ventricular failure in the presence of massive peripheral edema. *Am J Cardiol* 1980;45:841–847.
43. Abrams JA. The mystery of nitrate resistance. *J Am Coll Cardiol* 1991;68:1393–1396.
44. Armstrong PW, Armstrong JA, Marks GS. Pharmacokinetic-hemodynamic studies of intravenous nitroglycerin in congestive cardiac failure. *Circulation* 1980;62:160–166.
45. Elkayam U, Mehra A, Shotan A, Ostrzega E. Nitrate resistance and tolerance: Potential limitations in the treatment of congestive heart failure. *Am J Cardiol* 1992;70:98B–104B.

INDEX